The Atmosphere: A Very Short Introduction

VERY SHORT INTRODUCTIONS are for anyone wanting a stimulating and accessible way into a new subject. They are written by experts, and have been translated into more than 45 different languages.

The Series began in 1995, and now covers a wide variety of topics in every discipline. The VSI library now contains over 500 volumes—a Very Short Introduction to everything from Psychology and Philosophy of Science to American History and Relativity—and continues to grow in every subject area.

Very Short Introductions available now:

ACCOUNTING Christopher Nobes
ADOLESCENCE Peter K. Smith
ADVERTISING Winston Fletcher
AFRICAN AMERICAN RELIGION
Eddie S. Glaude Jr
AFRICAN HISTORY John Parker and
Richard Rathbone
AFRICAN RELIGIONS Jacob K. Olupona
AGEING Nancy A. Pachana
AGNOSTICISM Robin Le Poidevin
AGRICULTURE Paul Brassley and
Richard Soffe
ALEXANDER THE GREAT
Hugh Bowden
ALGEBRA Peter M. Higgins
AMERICAN HISTORY Paul S. Boyer
AMERICAN IMMIGRATION
David A. Gerber
AMERICAN LEGAL HISTORY
G. Edward White
AMERICAN POLITICAL HISTORY
Donald Critchlow
AMERICAN POLITICAL PARTIES
AND ELECTIONS L. Sandy Maisel
AMERICAN POLITICS
Richard M. Valelly
THE AMERICAN PRESIDENCY
Charles O. Jones
THE AMERICAN REVOLUTION
Robert J. Allison
AMERICAN SLAVERY
Heather Andrea Williams
THE AMERICAN WEST Stephen Aron
AMERICAN WOMEN'S HISTORY
Susan Ware

ANAESTHESIA Aidan O'Donnell
ANARCHISM Colin Ward
ANCIENT ASSYRIA Karen Radner
ANCIENT EGYPT Ian Shaw
ANCIENT EGYPTIAN ART AND
ARCHITECTURE Christina Riggs
ANCIENT GREECE Paul Cartledge
THE ANCIENT NEAR EAST
Amanda H. Podany
ANCIENT PHILOSOPHY Julia Annas
ANCIENT WARFARE Harry Sidebottom
ANGELS David Albert Jones
ANGLICANISM Mark Chapman
THE ANGLO-SAXON AGE John Blair
ANIMAL BEHAVIOUR
Tristram D. Wyatt
THE ANIMAL KINGDOM
Peter Holland
ANIMAL RIGHTS David DeGrazia
THE ANTARCTIC Klaus Dodds
ANTISEMITISM Steven Beller
ANXIETY Daniel Freeman and
Jason Freeman
THE APOCRYPHAL GOSPELS
Paul Foster
ARCHAEOLOGY Paul Bahn
ARCHITECTURE Andrew Ballantyne
ARISTOCRACY William Doyle
ARISTOTLE Jonathan Barnes
ART HISTORY Dana Arnold
ART THEORY Cynthia Freeland
ASIAN AMERICAN HISTORY
Madeline Y. Hsu
ASTROBIOLOGY David C. Catling
ASTROPHYSICS James Binney

Available soon:

For more information visit our website
www.oup.com/vsi/

Paul I. Palmer

THE ATMOSPHERE

A Very Short Introduction

OXFORD
UNIVERSITY PRESS

OXFORD

UNIVERSITY PRESS

Great Clarendon Street, Oxford, OX2 6DP,
United Kingdom

Oxford University Press is a department of the University of Oxford.
It furthers the University's objective of excellence in research, scholarship,
and education by publishing worldwide. Oxford is a registered trade mark of
Oxford University Press in the UK and in certain other countries

© Paul I. Palmer 2017

The moral rights of the author have been asserted

First edition published in 2017

Published in the United States of America by Oxford University Press
198 Madison Avenue, New York, NY 10016, United States of America

British Library Cataloguing in Publication Data

Data available

Library of Congress Control Number: 2016955253

ISBN 978-0-19-872203-8

Printed and bound by
CPI Group (UK) Ltd, Croydon, CR0 4YY

Contents

Acknowledgements

A big thanks goes to Alan Blyth, Massimo Bollasina, Hartmut Bösch, Martyn Chipperfield, Tim Garrett, and Ken Rice who suffered from reading early versions of individual chapters; I also thank two anonymous reviewers who provided thoughtful comments on the complete draft and made it much better. Thanks also go to Jenny Nugee and Latha Menon at OUP who have exhibited tremendous amounts of patience with me. The remaining errors in the book are mine alone.

For always being there, much love goes to Adèle, Lily, James, Missy, and Coco.

Finally, I will forever be indebted to Lilian Monks, J. O. Carter, John Barnett, and Daniel Jacob for providing me with opportunities.

List of abbreviations

Abbreviation	Definition
AU	Astronomical Unit describes the distance between Earth and Sun (value = 150 million km)
CFC	Chlorofluorocarbon
ELVES	Emission of Light and Very low frequency perturbations due to Electromagnetic pulse Sources
ENSO	El Niño Southern Oscillation
ESA	European Space Agency
HCFC	Hydrochlorofluorocarbons
ISS	International Space Station
ITCZ	InterTropical Convergence Zone
MAPS	Measurement of Air Pollution from Satellite
MAVEN	Mars Atmosphere and Volatile EvolutioN
MCF	Methylchloroform
MJO	Madden–Julian Oscillation
MOPITT	Measurements Of Pollution in The Troposphere
NASA	National Aeronautics and Space Administration
NOAA	National Oceanic and Atmospheric Administration
Pg	Peta-gram (value = 10^{15} g)
ppb	Parts per billion (value = 10^{-9})
ppm	Parts per million (value = 10^{-6})
PSC	Polar Stratospheric Cloud
Tg	Tera-gram (value = 10^{12} g)
TOMS	Total Ozone Mapping Spectrometer
UAV	Unmanned Airborne Vehicle
UT	Universal Time

List of illustrations

Chapter 1
What is special about Earth's atmosphere?

Which of us haven't occasionally been entranced by the perpetual movement of clouds or the majesty of lightning? Who hasn't wondered why the sky is blue in the daytime, dark at night, and sometimes an explosion of colour at sunset? Or why most of Earth's rainforests are located near the equator while most major deserts are located at thirty degrees north and south? But how many of us have pondered why Earth's atmosphere is the way it is, what makes it special, and appreciated its many chemical and physical properties that we rely on to survive?

Even though we can routinely gaze through Earth's atmosphere to the Sun, Moon, or twinkling stars, we only experience a small portion of it. Burj Khalifa in Dubai, the world's tallest building, rises almost 830 m, still less than 1 per cent of the height of the atmosphere. At the peak of Mount Everest, Earth's tallest mountain, we are only 9 km into our vertical exploration of the atmosphere. But by the time we had visited an altitude of 2.4 km, easily reached in the Pyrenees or Rocky Mountains, we would already very likely to be suffering from sickness due to the lack of oxygen. Further ascent requires specialist training and apparatus and therefore is limited to the few.

Despite this physical hurdle, many people reach heights just over 12 km in a pressurized metal container on a long-haul flight, but this altitude is still only a fraction of the vertical range of our atmosphere.

Man-made objects, on which we often rely on a daily basis, reside at much higher altitudes. Weather balloons, regularly launched around the globe, collect data up to 30 km. The International Space Station, which is home for at least three people, orbits Earth every 90 minutes at an altitude of 400 km. The Global Positioning System satellites, that we all know and love, orbit Earth at an altitude of 20,000 km above the surface. Admittedly, at the altitudes of man-made satellites the atmosphere is to all practical purposes non-existent.

While few humans have travelled up through every atmospheric layer, we all feel the full weight of it as it bears down on the surface. The mass of the Earth's atmosphere is approximately 5,200,000,000,000,000,000 kg (or in scientific shorthand 5.2×10^{18} kg), a mind-boggling number. Although this estimate is based on a detailed mathematical calculation you can almost get there by combining the surface pressure (often quoted in regional weather forecasts) with the surface area of the Earth. To put this number into context it is approximately 10 million times the weight of the human population, after assuming a generous average individual weight for a person; although if we continue with current trends my generosity may be less obvious in the future.

In the simplest sense, an atmosphere is a diffuse fluid that represents a very thin envelope around our planet—compare the height of the atmosphere (some 600 km) with the radius of the Earth (6,371 km). There are two common definitions for the height of the atmosphere taken from different disciplines.

Aeronautical engineers use the Kármán line, named after Hungarian-US scientist Theodore von Kármán (1881–1963), which is around 100 km above mean sea level and is defined as the boundary between the atmosphere and outer space. Travel beyond here and you are thought to have gone far enough to be awarded astronaut wings. Although there is no abrupt boundary in the atmosphere, it does get progressively less dense with altitude, with the Kármán line denoting the point where the atmosphere is too thin to support aeronautical lift.

Physicists and planetary scientists tend to use the exobase, which is the bottom of the exosphere some 600 km above Earth, mainly because it relies on thermodynamics for its definition. Particles are still bound by gravity but the density of particles is too low for them to behave like a gas and collide with each other. The advantage of this definition is that it can be readily applied to other planets where the exobase can be less than a few metres in depth.

Earth's atmosphere is generally considered as several interconnected layers (Figure 1), which have different characteristics mainly determined by the density of air and their relative proximity to Earth's surface and outer space. The troposphere describes the lower atmosphere from the surface to the tropopause at 10–15 km. The stratosphere describes the layer between the tropopause and the stratopause at 50 km. Above the stratopause lies the mesosphere, which is the layer stretching from the stratopause to the mesopause at 100 km. The stratosphere and mesosphere are collectively known as the middle atmosphere. The thermosphere lies above the mesosphere and takes us to 500–1000 km. Above the thermosphere is the exosphere, which describes the layer from the thermopause (exobase) to the near vacuum of outer space. The thermosphere and exosphere are collectively known as the upper atmosphere. The ionosphere describes the region that straddles the middle and upper atmosphere, which is ionized by solar ultraviolet radiation. We will explore more about each atmospheric layer later in this book.

Earth's atmosphere receives heat and light (different forms of radiation) from our star. The Sun continues to play a central role

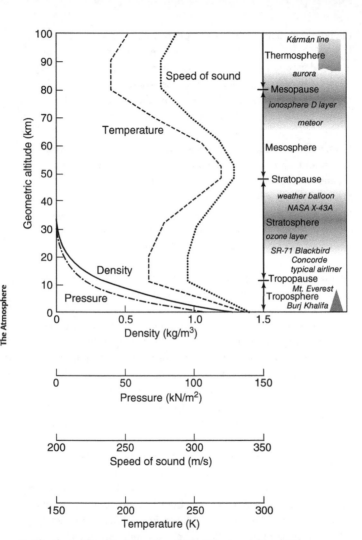

1. The thermal vertical structure of Earth's atmosphere, with associated landmarks.

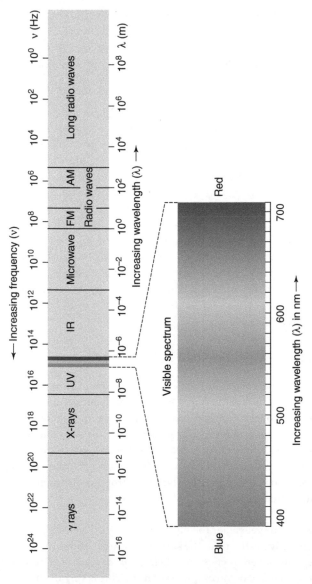

2. The electromagnetic spectrum describes the type of radiation associated with different wavelengths.

in shaping Earth's atmosphere and the atmospheres (or their absence) of other solar system planets. Technically speaking, our sun is a low-mass main sequence star that is halfway through its life. Our star emits electromagnetic radiation with a range of wavelengths (Figure 2).

Human rod and cone cells in the retina have peak sensitivity at visible wavelengths (fractions of a millionth of a metre), which represent only a small portion of the available spectrum. For most of Earth's atmosphere, radiation emitted at ultraviolet (a few billionths of a metre) and infrared (a few millionths of a metre) wavelengths is the most relevant. Our atmosphere is largely transparent to incoming solar radiation. About half of this radiation is absorbed by Earth's surface after which it is transformed into heat and radiated as infrared radiation back into the atmosphere. Atmospheric gases interact with radiation at these longer infrared wavelengths so not all of it escapes. This is called the greenhouse effect, although, as we discuss later, this is a misnomer because the atmosphere does not really behave like a greenhouse.

The lower atmosphere

The lowest part of the atmosphere, the troposphere, stretches from Earth's surface to around 10–15 km and represents about 75 per cent of atmospheric mass. Within the troposphere, atmospheric temperature decreases by –6.5°C for every kilometre increase in altitude until we reach the tropopause, where temperatures are approximately –50°C. Because of surface heating the tropopause is higher at low latitudes (tropics) and lower at higher latitudes (poles).

Why does air get cooler the higher up you go? The atmosphere is effectively heated from the surface upwards. And while the lower atmosphere is constantly being mixed (tropos is Greek for turning), air is a poor conductor of heat. This is why we are encouraged to wear layers in cold weather—air trapped between these layers slows down the loss of heat from your body to the outside. The troposphere, bounded by the tropopause, encompasses the weather we experience. It is also where the land and ocean meet the atmosphere and where you and I breathe air.

The air we breathe comprises 78 per cent nitrogen, 21 per cent oxygen, with everything else being trace gases. As we will see in Chapter 4 it is that 1 per cent of trace gases in the atmosphere that makes life on Earth interesting.

The middle atmosphere

The temperature minimum at the tropopause acts as a barrier to water being transported into the stratosphere, where it can affect stratospheric temperatures, ozone, and surface climate. Directly above the tropopause, atmospheric temperature increases with height up until the stratopause at approximately 50 km, where atmosphere temperatures can exceed 0°C.

Within the stratosphere, there is a natural level of ozone that is determined by ultraviolet light interacting with molecular and atomic oxygen. Heating of the stratosphere is mainly due to ozone absorbing incoming ultraviolet and visible solar radiation, with smaller contributions due to carbon dioxide and water vapour absorbing at shorter infrared wavelengths. The stratosphere contains only 10 to 20 per cent of the total atmospheric mass, but changes in stratospheric composition are important because they affect the balance of incoming and outgoing radiation.

There are at least two interesting human endurance milestones within the stratosphere. Between 18 and 20 km lies the Armstrong limit that corresponds to an atmospheric pressure at which water boils at body temperatures; flying at or above this limit without the aid of a pressurized suit or container would literally boil off your insides. Aviators have worn pressure suits since the mid-1930s to avoid hypoxia at high altitudes, where a reduction in the partial pressure of oxygen within the body affects brain functioning. For a while in the 1940s, this hard limit precluded aircraft observations of the atmosphere far into the stratosphere. The second milestone is the current record for a freefall jump, on 24 October 2014, held by Google Vice President Alan Eustace, which is just shy of 41.5 km. The atmosphere is so thin at these altitudes that falling does not produce much of a sound. No air, no air resistance.

Above the stratosphere is the mesosphere. Together the stratosphere and mesosphere are called the middle atmosphere. The depth of the mesosphere is 40–50 km, ranging from the stratopause to the mesopause at 100 km. Temperature decreases rapidly with altitude in the mesosphere. This is due to decreasing solar heating because of progressively fewer molecules available to absorb at these wavelengths. Another factor is cooling from carbon dioxide, which absorbs energy from scattered light and releases energy in all directions so that a portion of it will be carried away from the mesosphere. The mesopause is the coldest part of the atmosphere, with a temperature of less than –100°C. Scientists use instruments on small rockets to sample the mesosphere but rocket launches are few and far between, with the result that we still know very little about this exotic region of the atmosphere.

The mesosphere is home to the highest clouds in the atmosphere: noctilucent clouds. These clouds form directly from ice particles (less than ten millionth of a metre across) at high latitudes during summer months when temperature fall below –120°C, as opposed to clouds in the lower atmosphere that form when water collects on small particles. When these ice crystals grow large enough to scatter incoming solar radiation they become visible from the ground during twilight, when the Sun is below the horizon but light is being scattered in the upper atmosphere.

Recent scientific studies have linked the increased frequency of the noctilucent clouds as a harbinger for changes in climate, but the jury is still out. The mesosphere is also home to electrical phenomena with names such as red sprites and ELVES that would not be out of place in Tolkien's Middle Earth; we will return to these later on.

Meteoroids are small rocky or metallic chunks, often no bigger than a grain of sand you might find on a beach, that travel through space. They originate from disintegrating comets and asteroids. The vast majority of meteoroids disintegrate in the mesosphere. Once they enter Earth's atmosphere at a speed of some 20 km/s they experience substantial aerodynamic drag that heats up the material to incandescent temperatures. The result is a glowing object with a trail of glowing particles in its wake. This is commonly called a meteor or shooting star. Meteors typically occur in the mesosphere. The origin of the word meteor is from the Greek meteōros that means 'high in the air'.

The upper atmosphere

The thermosphere lies above the mesosphere and the exosphere lies atop that; collectively they are known as the upper atmosphere. We are now at altitudes where only astronauts and artificial satellites reside. Above the mesopause the atmosphere can be characterized not only by temperature (as it is lower down) but also by electric conductivity or electron density. Above the mesosphere, gases in the atmosphere start to stratify according to their molecular mass, with the lower mass gases diffusing higher up.

The thermosphere stretches from about 100 km to 500–1000 km above Earth's surface. Temperatures increase dramatically in the lower thermosphere (up to 300 km) from the lowest atmospheric temperature at the mesopause to over 1000°C and then remain relatively constant. The highest temperatures are highly dependent on the level of solar activity. The heat source for the thermosphere is from the absorption of very high-energy particles of light (photons, usually at ultraviolet and X-ray wavelengths), by nitrogen and oxygen (as molecules and single atoms), and from collision with high-energy particles in the Van Allen radiation belts (a layer of charged particles held in place by Earth's magnetic field). These high-energy particles cause the atmosphere to ionize, i.e. gain or lose electrons to form ions.

The progressively decreasing density of gases within the thermosphere leads to two new phenomena. Even though thermospheric temperatures are extremely high you would still be very cold (if you were able to survive at this altitude). The transfer of heat needs a medium through which to travel and in the absence of air the energy lost by your thermal radiation would far exceed the energy acquired by direct contact with the dilute gases. By the time you reach 160 km you will have entered the anacoustic zone where the density is too low to support the transmission of sound audible to the human ear. In space no one can hear you scream about how cold you are.

The International Space Station (ISS) orbits in the thermosphere at about 400 km above Earth's surface and travels at a speed of nearly 8 km/s so that the occupants orbit the Earth every 90 minutes. One of the objectives of the ISS is to act as a test bed for satellite equipment (Chapter 5) and a number of international space agencies have installed instruments that can view the atmosphere and the land surface.

Finally, we reach the exosphere. It extends from the exobase (or thermopause) to the near vacuum that is outer space. At altitudes within the exosphere, where the density of neutral gases is extremely low, the concept of a gas falls apart. The distance covered by a gas molecule between collisions is sufficient large (hundreds of kilometres) that it can accelerate in excess of the speed that is necessary to escape Earth's gravitational pull. This process is called Jeans escape and is partly responsible for the leakage of hydrogen from Earth's atmosphere, which is only important on geological timescales. All unmanned Earth-orbiting satellites, e.g. Meteosat for studying weather and the Global Positioning System for navigation, reside in the exosphere.

The ionosphere describes the region of the middle and upper atmosphere that is ionized by solar ultraviolet radiation: the charged atmosphere. It spans from the upper mesosphere to the exosphere. The amount of ionization depends on variations in solar activity, and therefore exhibits diurnal, geographical, and seasonal variations. The importance of the ionosphere lies in its ability to reflect electromagnetic radiation. We relied on this property for long-distance communication before the era of communication satellites. A single 'hop' off the ionosphere can help transmit a radio signal a distance of up to 3,500 km, well beyond the horizon; transatlantic connection can be obtained within two or three hops. Nowadays, we use orbiting satellites to reflect information across the globe.

Because of its importance in the transmission of electromagnetic radiation, understanding, and thereby predicting, the variability of the ionosphere (so-called space weather) is a current research question.

While the thermosphere is cloudless and free of water vapour, the neutral and charged atmosphere can produce awe-inspiring phenomena in the form of auroras that occur at northern (borealis) and southern (australis) high latitudes. They arise as a result of ionized gases and neutral gases, altered by incoming high-energy particles, which emit light in the process of returning to their normal (ground) state. The colour of the aurora depends on the gases emitting the photons: green and orange-red is produced by oxygen and blue or red is produced by nitrogen. The swirling patterns of the aurora are due to the incoming electrons being trapped and spiralling within Earth's magnetic field.

Table 1 Mean characteristics of solar system planets

	Mass (relative to Earth)	Mean radius (relative to Earth)	Mean distance from Sun (AU)	Dominant composition
Sun	333,000	109.3	—	H_2, He
Mercury	0.0553	0.383	0.387	H_2, He, O_2
Venus	0.815	0.950	0.722	CO_2, N_2
Earth	1	1	1.000	N_2, O_2
Mars	0.107	0.532	1.520	CO_2, Ar, N_2
Jupiter	317.83	10.97	5.200	H_2, He
Saturn	95.159	9,14	9.580	H_2, He
Uranus	14.536	3.98	19.20	H_2, He
Neptune	17.147	3.86	30.10	H_2,He

Solar system planetary atmospheres

The Earth is one of eight planets in the Solar system. In order from the Sun they are: Mercury, Venus, Earth, Mars, Jupiter, Saturn, Uranus, and Neptune (Table 1). The four nearest planets are called terrestrial planets as they have a rocky surface. The outer four planets are gas or ice giant planets. The distance between the Earth and Sun is 150 million km, which is often referred to as one Astronomical Unit (abbreviated as AU). Table 1 shows the basic attributes of the Sun and its planets.

Mercury is too hot and small to sustain a substantial atmosphere. Its exosphere is very close to its surface and comprises mostly elements that originate from the solar wind (a continuous stream of charged particles emitted by the Sun), meteor impacts, and the breakdown from the crustal surface. Venus has a mass and size similar to Earth so it has sufficient gravity to retain an atmosphere. Venus' atmosphere is much hotter than Earth's so it has already lost hydrogen via various atmospheric escape mechanisms and can barely contain the next light element helium. The atmosphere of Venus is dominated by carbon dioxide (96.5 per cent) with a minor contribution from nitrogen (3.5 per cent). The enormous amount of atmospheric carbon dioxide results in a surface temperature in excess of 470°C, which makes Venus the hottest planet in the solar system. Mars has a small mass and volume compared to Earth but it is able to support a thin atmosphere, which is mostly carbon dioxide (95.97 per cent) with small contributions from nitrogen (1.93 per cent) and argon (1.89 per cent). Any abundant atmosphere it once had has since been stripped away by solar winds.

The four outer planets, Jupiter, Saturn, Uranus, and Neptune, all have atmospheres dominated by hydrogen and helium. Gas phase atmospheres blend into liquid interiors once the atmosphere pressure increases beyond a critical point determined by thermodynamics. Consequently, there is no clear boundary between the atmosphere and the surface. The abundance of helium increases with distance from the Sun.

Stars like the Sun form in giant molecular clouds, which are comprised mostly of hydrogen and helium but also contain a small amount of dust grains. Conservation of angular momentum means that this material cannot fall directly into the central proto-star but instead forms a proto-stellar (sometimes called a proto-planetary) disc. This disc provides a mechanism for transporting angular momentum outwards, allowing mass to flow onto the central proto-star, and is also a site for planet formation. The dust grains grow initially via collisions. Those that become sufficiently massive can then gravitationally attract other nearby grains, ultimately growing to become kilometre-sized planetesimals. In the inner regions of the discs, these objects continue growing to finally form rocky, terrestrial planets.

Beyond the snowline (or iceline)—at about 3 AU in a disc around a young Sun-like star—hydrogen compounds condense onto these dust grains, providing more solid mass and allowing the objects in the outer parts to grow more rapidly and become more massive that those in the inner regions. These outer objects become the cores of gas giant planets like Jupiter and Saturn. The recent discovery of Jupiter-mass planets, outside of our solar system (described later), which orbit much closer to their host star have challenged our understanding of planet formation. Most scientists still think it is most likely that these gas giants formed far from their star and then migrated inwards.

The habitable zone in the solar system describes the range of distances away from the Sun that could sustain liquid water at the surface. A simple calculation gives the potential of a planet to be habitable; however, a planet lying beyond this zone does not preclude it being habitable. The inner edge of the zone (closest to the Sun) is defined as being where a planet is sufficiently hot that it can develop a wet middle and upper stratosphere. At these altitudes, photons are energetic enough to break apart the strong bond between hydrogen and oxygen, thereby allowing the hydrogen to eventually escape from the atmosphere. The remaining oxygen is left to oxidize the planet's crust. This process effectively represents a loss of water from the planet. The outer edge of the habitable zone (furthest from the Sun) is defined where a planet is too cold to maintain liquid water.

The inner edge is determined mainly by the distance from the Sun, and is closer to our present position than you might have expected. On Earth at 1 AU, the annual mean global flux (rate of flow per unit time and space) of solar electromagnetic radiation is 1,366 watts per square metre. This flux decreases as an inverse square law so that the radiation received at 2 AU would be $(1 \text{ AU}/2 \text{ AU})^2$ or 25 per cent that received at 1 AU. On Venus, at 0.72 AU, the radiation is $(1 \text{ AU}/0.72 \text{ AU})^2$ or a factor of 2.8 higher than that received on Earth. The resulting surface temperature would be greater than 40°C, which is sufficiently hot to allow large amounts of water vapour to accumulate in the atmosphere. The resulting warming of the atmosphere and surface via the greenhouse effect forms a feedback mechanism so that the water can never condense into liquid or ice and is eventually lost to space. The inner edge of the habitable zone is at 0.95 AU from the Sun. This corresponds to a modest 10 per cent increase in solar radiation currently received on Earth.

The outer edge is a bit harder to define but we can use Mars (at 1.52 AU) as a useful reference point from which to do so. At the distance of this planet the solar radiation is $(1AU/1.52 AU)^2$ or 43 per cent that received on Earth. Numerical calculations estimate that if the Sun emitted 85 per cent less radiation than at present (as it might have done when it was much younger) it would be impossible to maintain any liquid water on the surface of Mars. This fraction of the solar radiation received on Earth today ($0.85 \times 0.43 = 0.37$) effectively defines the outer edge of the zone to be 1.65 AU. Interestingly, given solar insolation, Mars is within a potential habitable zone.

Both the calculations of the inner and outer edges are subject to large uncertainties. Clouds will likely increase the planetary reflectance as the planet warms and that will act to shrink the inner edge inwards towards the Earth. The outer edge could easily be pushed outwards if carbon dioxide ice clouds or greenhouse gases can warm the planet. Despite these uncertainties and gaps in knowledge, the calculations illustrate the main point of the habitable zone. Earth is within this zone. It is cooler than Venus and warmer than Mars. It also means that enough water vapour could accumulate in the atmosphere and establish a greenhouse effect before condensing into oceans. The contrasting atmospheric fates of Venus, Earth, and Mars are sometimes called the Goldilocks Principle because Earth evolved at just the right location from the Sun, with the alternative locations being too hot (Venus) or too cold (Mars). The small probability of this ideal positioning occurring elsewhere is used to form a counter-argument to finding life on other planets, but this is compensated for by the staggering number of planets in galaxies in the Universe. This counter-argument also assumes that liquid water is required to sustain life, which may not be valid.

Exoplanets

Extrasolar planets are planets orbiting stars outside the solar system. Based on the Copernican principle, the assumption is that planets orbiting other stars should not behave any differently to those in the solar system so that we could apply this knowledge to any stellar system. Studying exoplanets is a relatively new field, reflecting the substantial measurement challenges that still have not been fully addressed by current instruments.

It was not until 1995 when the exoplanet field was blown open with the discovery of a Jupiter-mass planet that was orbiting the star 51 Pegasi. What was astonishing about this discovery was that the planet was orbiting only 0.05 AU away from the star. The close proximity of this large planet could then not be explained by models of giant planet formation, and even now theorists are debating how it could be possible. Since the 1995 discovery many hot Jupiters have been discovered.

Most current theories suggest that these bodies were formed further away from their host star and either migrated inwards through interactions with the disc from which they were forming, or were later scattered inwards by other planets in the same system or by stellar companions. As we have seen with the solar system planets, the formation and evolution of a planetary mass has implications for its atmosphere.

Most of the excitement that surrounds exoplanet discovery science is from the possibility that we might find another Earth that is orbiting a star that is similar to our Sun so that it could (or does!)

support life of sorts. This is a young and fast-moving scientific field and while a few candidate exoplanets have already been identified that have similar temperatures to Earth there will undoubtedly be more out there. Other planets that have been found in potentially habitable zones are much larger than Earth (super Earths) and orbit around stars that are much smaller than the Sun. These small stars are less luminous so the habitable zone is closer to the star so the planets are typically tidally locked—much like the Moon is to the Earth. This will have implications for the movement of the atmosphere. The peak wavelengths emitted by these stars are different from the Sun, which may have implications for photosynthesis and subsequently for atmospheric composition. To understand more about whether these exoplanets can sustain life we have to understand how the planet was formed and whether (and where) it lies in the habitable zone of the host star.

Let us assume for the moment that we have found a planet not unlike Earth that is orbiting a star similar to the Sun. We think it can potentially support life, but does it? We can't readily visit this planet using current modes of space travel so we have to rely on remote spectroscopic observations that look at light that is reflected from the planet at different wavelengths. Spectroscopic measurements describe the interaction between radiation and matter. Because gases absorb or emit at particular wavelengths and atmospheric aerosols affect observed spectra in a broadly predictable manner, we use these data to determine how much of a gas is in the atmosphere. This is described further in Chapter 2. We are currently at the very limit of our capabilities with only the first tantalizing glimpses of what makes up the atmospheres of these exoplanets. But by the mid-2020s after the launch of the James Webb Space Telescope we should be in much better shape.

Using these spectroscopic signatures we can look for atmospheric or land-based biosignatures, unique fingerprints of biological life. But what do we measure and how do we measure it? One approach scientists have taken to address these questions is to reverse the telescope to view Earth from afar to understand what makes it unique. What measurements of Earth's atmosphere can we collect that definitively identify life? It is worth reminding ourselves that life on other planets, even if it has been subjected to similar environmental conditions, is unlikely to have followed the same evolutionary path as on Earth. With that in mind, biosignatures on Earth may not directly relate to those we might observe from other planets.

On Earth the presence of atmospheric oxygen is a sign of life and via oxidation so is atmospheric ozone. Recent work showed that carbon dioxide can photo-disassociate into carbon and oxygen so that a significant non-biological source of oxygen might exist for atmospheres dominated by carbon dioxide under certain stellar radiation environments. While they are flawed, oxygen and ozone remain our best biosignatures in the context of additional information. Atmospheric water vapour is essential to being in the habitable zone but it is not necessarily a signature of life. Methane on Earth has several biological sources but there is also a non-biological source from inorganic processes within the Earth's interior.

Other measures of life include carbohydrates and chlorophyll that can potentially leave a fingerprint on spectroscopic measurements. Chlorophyll is a green pigment found in biological material. It reflects radiation at visible wavelengths up until the red values when it becomes transparent. This is called the red edge and has been used as an indicator of life on Earth. At different wavelengths, chlorophyll can exhibit fluorescence where it absorbs and then re-emits light. Mineral and biological material both have fluorescent properties but Earth-orbiting satellites have shown they can reliably observe chlorophyll fluorescence, related to photosynthesis, at characteristic wavelengths. Another spectroscopic property is chirality, which describes, in our case, the molecular property to exist in two mirror-image states. This property would effectively depolarize signals that could be detected from spectroscopic measurements. Molecules such as sugars in the atmosphere could be linked with biological processes. The spectroscopic approach is routinely used to study Earth's atmosphere but current exoplanet technology that needs to overcome the challenges associated with observing and interpreting a small signal is still a way off. Even if we found any of the above biosignatures in the observed atmospheric spectra of an exoplanet, none would provide unequivocal evidence for life.

Other scientists have put forward the idea of using man-made chemicals (such as chlorofluorocarbons, discussed Chapter 4) as a marker for intelligent life. This neatly sidesteps the issue of placing the Earth-centric biological constraints on the search for extra-terrestrial life. It does, however, assume that alien life developed the need for similar technologies that we did, which is not impossible but unlikely. But if we reversed the telescope it would allow alien life to view us from afar and to identify synthetic chemicals in the atmosphere of the watery third rock from the Sun.

Chapter 2
Atmospheric physics

The evolution of Earth's atmosphere is tied closely with the Sun. The Sun is approximately halfway through the main sequence after which it will have exhausted its supply of hydrogen fuel and will grow into a red giant, and on the timescale of billions of years start to engulf some of the inner planets in the solar system. Even before this stage, the luminosity of the Sun will steadily increase until Earth will no longer be capable of supporting life, as we know it.

Earth's radiation budget

The Sun emits electromagnetic radiation in all directions at a wide range of wavelengths. The rate at which solar flux falls onto a surface per unit area and per unit time, irradiance, reduces as an inverse square law so that for every 2 km from the Sun, solar irradiance drops off by a quarter. By 1 AU the solar radiation intercepted by the sunlit Earth disk (Figure 3) is averaged over a year 340 W/m^2, which is thankfully only 0.05 per cent of what is emitted by the Sun. Because Earth is (approximately) spherical and inclined, it receives less solar energy per unit area at higher latitudes: the midlatitudes receive about 70 per cent of the energy compared to the equator and the polar regions receive about 40 per cent of the equatorial energy. As we discuss in Chapter 3 this establishes an energy

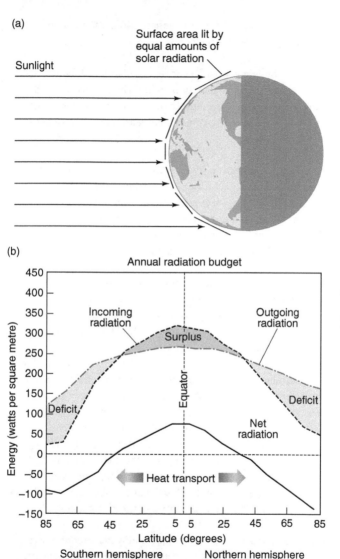

3. Solar energy intercepted by Earth (a) is spread across a larger geographical area at higher latitudes, which helps to (b) establish a surplus of energy at low latitudes and a deficit of energy at high latitudes.

imbalance, which results in atmospheric circulation to redistribute the heat energy towards the poles.

A radiation budget is a useful concept to distinguish between incoming and outgoing radiation, equilibrium between these terms, and departures from this equilibrium due to, for example, increasing levels of clouds, greenhouse gases, and atmospheric aerosols (Figure 4).

The incoming radiation to Earth's atmosphere comprises visible wavelengths and contributions from infrared (longer) and ultraviolet (shorter) wavelengths; these constitute shortwave radiation. Solar irradiance at Earth's surface (insolation) peaks at visible wavelengths because the atmosphere is essentially transparent to light at these wavelengths.

The higher energy ultraviolet photons are mostly absorbed in the middle and upper atmosphere, which effectively shields us from radiation that is more likely to damage human and plant cells. The lower-energy infrared photons are mostly absorbed by water in the lower atmosphere.

The atmosphere absorbs approximately a quarter of the incoming solar radiation. As we will shall see in Chapter 4, only certain atmospheric gases absorb and emit radiation, and only at specific wavelengths. The ability of a molecule to absorb radiation is associated with its ability to move (e.g. vibrate, bend) and accelerate, which can cause changes in the distribution of its charge. Symmetric molecules like nitrogen (N_2) and oxygen (O_2), which represent 99 per cent of Earth's atmosphere, cannot change their charge distribution irrespective of their suppleness.

In contrast, gases like carbon dioxide and water vapour have a whole range of movements that can result in a dipole moment, which is a measure of the molecular distribution of charge. These molecules strongly absorb infrared radiation.

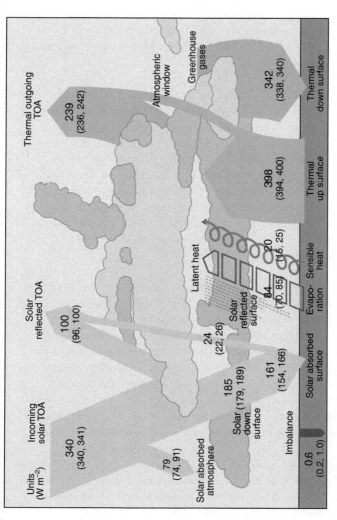

4. Overview of Earth's radiation budget. Numbers in parentheses denote the range of values that is consistent with observations.

Absorbed radiation will cause atmospheric molecules to reach an excited state that they cannot sustain indefinitely and eventually they emit radiation that allows them to return to their unexcited state. This radiation is absorbed by the atmosphere again, absorbed by Earth's surface, or is returned to space. We will return to clouds later.

Approximately a third of all the 340 W/m^2 of incoming radiation is reflected back to space. This reflection is usually described by an albedo where a larger fractional value denotes a higher reflectivity. Every surface reflects radiation to some degree, depending on the wavelength of the incident light so that a reflective surface for visible light will not necessarily be reflective for shorter or longer wavelengths. For example, snow can be a perfectly reflective surface for visible light but is almost perfectly absorbing (black) at longer infrared wavelengths. Snow is a very complicated material because its albedo also varies with a number of other factors, including age, wetness, and depth.

About half of the incoming sunlight is absorbed by Earth's surface. The Earth subsequently warms up with this absorbed radiation. Anything that has a temperature above absolute zero (−273.15°C) radiates energy and since nothing can be absolute zero everything radiates energy; the Sun, Earth, you, and I all radiate energy. The differences in the energies radiated by these objects are due to their temperature and size. Scientists often refer to blackbody radiation, where the blackbody is an idealized object that absorbs incident radiation with 100 per cent efficiency and emits radiation at an equivalent rate, thereby maintaining a constant temperature.

To a first approximation Earth is a blackbody. So that as Earth's surface warms up it radiates thermal infrared radiation that has a longer wavelength than the incoming sunlight. Greenhouse gases absorb thermal infrared radiation and subsequently radiate some of this energy back to Earth and contribute to the habitable temperature of Earth's surface, and some of the energy towards space.

Heat is also transported from the surface by conduction and convection. Evaporation of water cools the surface and heats up the atmosphere when the vapour eventually condenses or ice particles form (discussed in the section 'Atmospheric water').

Transport of radiation through the atmosphere

The atmosphere interacts with solar radiation in a number of ways and this interaction is sometimes easier to understand when light is described as a particle (quantum physics) or as a wave (classical physics).

In general, light can be transmitted, absorbed, or scattered as it passes through the atmosphere. Figure 5 shows a schematic of these processes. Some fraction of light (a) will pass unimpeded through the atmosphere (transmission). Atmospheric gases will typically absorb some fraction of light either partially or completely (b), which we discuss in Chapter 4. Finally, some fraction of light (c) will be scattered by the atmosphere, so that the light path deviates away from the path that it would have taken in the absence of an atmosphere.

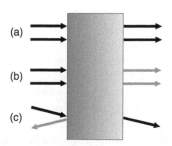

5. **The fate of light as it passes through an atmosphere.**

Rayleigh scattering, named after UK physicist John Strutt (1842–1919; 3rd Lord Rayleigh), describes elastic scattering of light. In Earth's atmosphere, Rayleigh scattering is due to atoms and gases, but also to atmospheric aerosols (small particles suspended in the atmosphere) that are much smaller than the wavelength of light. Elastic scattering means that the energy of light is the same before and after the collision, although the direction of movement of the particle may change after the collision. The reason why the atmosphere scatters light is related to its electromagnetic nature. Rayleigh scattering depends on the electric polarizability of molecules in the scattering medium. For molecules that are polarizable, incident light induces a dipole moment that can then subsequently emit radiation, which we see as scattered light. Scattering of the incident light by free, unbound charged molecules (known as Thomson scattering, after the UK physicist J. J. Thomson (1856–1940) who first explained this phenomenon) results in much weaker radiation. The intensity of Rayleigh scattering is extremely sensitive to the wavelength of light (being proportional to the inverse fourth power of the wavelength) so that shorter wavelengths of visible light (e.g. blue) are scattered more strongly than longer wavelengths (e.g. red): if the wavelength of light is halved Rayleigh scattering will increase by a factor of 16.

Rayleigh scattering is responsible for the blue colour of the daytime sky, as observed from the ground: blue light is scattered more efficiently in visible light. At sunrise or sunset when the Sun is just above the horizon, light directly from the Sun takes a longer path through the atmosphere so by the time an observer sees the light the shorter wavelength light has been scattered leaving the longer wavelength yellow and red colours. During twilight, the hours between dawn and sunrise or between sunset and dusk when the Sun is below the horizon, sunlight scatters off the upper atmosphere to illuminate the lower atmosphere. Ozone in the stratosphere is responsible for the deep blue colour of the twilight atmosphere because it absorbs red and orange light. Without ozone the twilight atmosphere would be a pale grey-yellow colour.

Mie scattering, named after German physicist Gustav Mie (1858–1957), is the more general form of elastic scattering in the atmosphere but is more typically associated with light scattering by particles that have a diameter similar to the wavelength of light being scattered. The scattering efficiency is directly proportional to the diameter of the particle and inversely proportional to the wavelength. Forms of Mie scattering include scatter from smoke, where water and atmospheric aerosol scatter incoming light. Scattering by clouds and atmospheric aerosols can, for example, improve the transport of light into forest canopies leading to more efficient photosynthesis.

Polarization is a hidden property of light when it is considered as a wave. Light is polarized if it oscillates in one plane (Figure 6). If we consider light that comprises several waves that are oscillating along different planes, there is no coherent direction to their electric and magnetic fields. This is called unpolarized light (Figure 6). Sunlight has a polarization only one part in a million. Many kinds of atmospheric particles and surfaces can polarize light. For instance, Rayleigh scattering can polarize sunlight. Light can also be polarized after being reflected off the ocean, or any surface such as a wet road, at a certain angle. We currently use knowledge of polarization for sunglasses by blocking out horizontally polarized light through the lenses that would otherwise cause glare. While we humans have some sensitivity to polarized light, animals can see polarized light sufficiently well to use it to their advantage: navigation, detecting water surfaces, and biological signalling.

Figure 7 shows the net result of solar radiation after it has passed through an atmosphere without clouds or aerosols. The Sun and atmospheric matter radiates at short wavelengths (ultraviolet and visible light), while Earth's surface and atmospheric

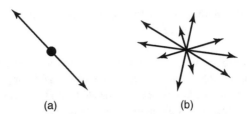

(a)   (b)

6. Polarization of electromagnetic radiation in the direction of propagation: (a) polarized light and (b) unpolarized light.

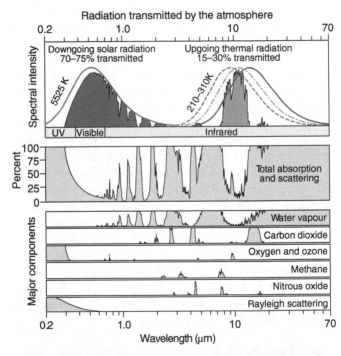

7. The Earth's solar radiation spectrum at ultraviolet, visible, and infrared wavelengths as observed at sea level. Blackbody curves at temperatures of 5525, 210, and 310 degrees Kelvin are superimposed to describe radiation from the Sun and the extremes of Earth's surface.

matter radiates at long wavelengths (infrared light) so even without the Sun, Earth's atmosphere would still be awash with radiation but at those longer wavelengths. At infrared wavelengths the main absorbers include water vapour, carbon dioxide, ozone, methane, and nitrous oxide. With the exception of ozone these gases mainly reside in the troposphere. In an atmospheric column most of the absorption of infrared radiation is due to water vapour with significant features from carbon dioxide and isolated features from other gases. Gases absorbing thermal (longer) infrared wavelengths are most sensitive to changes in the free troposphere above the lowest few kilometres of the atmosphere. Gases absorbing at shortwave infrared wavelengths are most sensitive to the lower troposphere. These gases also emit infrared radiation dependent on their temperature in the atmosphere.

The broad spectral region between 8 and 14 micrometres (10^{-6} m also known as microns) is called the atmospheric window because absorption is generally weak with the exception of the band near 9.6 microns associated with O_3. Within this window looking from space downwards we observe IR radiation from cloud tops or the land/ocean surfaces without it first being absorbed by the atmosphere. Details of this window vary with local environmental conditions such as water vapour and surface temperatures. Some atmospheric measurement instruments exploit these absorption features to infer non-water vapour trace gas concentrations in the lower atmosphere.

At ultraviolet and visible wavelengths the main absorbers are molecular oxygen, ozone, and water vapour. On closer inspection this spectral region is rich in information about many surface air pollutants that only weakly absorb at these wavelengths. As we discuss in Chapter 4, being able to extract the information from this spectral region heralded the start of space-based observations of atmospheric chemistry in the lower atmosphere.

Greenhouse effect

The greenhouse effect describes the absorption and emission of outgoing thermal infrared radiation by the atmosphere. The ability of the atmosphere to do this relies on greenhouse gases, which is a name given to gases that absorb and emit radiation at the relevant infrared wavelengths. Figure 7 shows that the principal gases that are active at these infrared wavelengths include water vapour, carbon dioxide, ozone, methane, nitrous oxide, and (not shown) chlorofluorocarbons. Broadly speaking, Earth's greenhouse effect is predominately due to water vapour with a substantial contribution from carbon dioxide. However, water vapour responds to changes in temperature, which are in turn controlled by atmospheric levels of non-condensable greenhouse gases such as carbon dioxide. The wide variety of sources and chemistry of these gases is discussed in Chapter 4. In the absence of greenhouse gases Earth's surface would be approximately 30°C cooler than it is now and would likely be covered in snow and ice. The modern-day concern about greenhouse gases is associated with the human-driven contribution that is now much higher than any previously reported natural levels, and continues to increase at a much faster rate than could be supported by any natural process.

Joseph Fourier, the French mathematician and physicist (1768–1830), famous for the Fourier series, knew that heated surfaces emitted radiation. The Earth was warmed by solar radiation and cooled by emitting thermal radiation back into space so why was Earth not a frozen planet? He reasoned that the atmosphere was acting as a blanket and keeping Earth's surface warm. He then likened the situation to a box with a glass cover. When the box was exposed to sunlight the interior of the box heated up after contact with the surface while the glass prevented the warmed air from escaping. These are different processes to the ones in the atmosphere, which involve the convection of heat and the absorption and emission of infrared radiation. However, it was an accessible analogy that is still widely adopted.

A few decades later John Tyndall, a British physicist (1820–93), measured in his laboratory the thermal absorption properties of water vapour, carbon dioxide, and other atmospheric constituents thereby supporting the idea put forward by Fourier. It was also Tyndall who took a keen interest in the waning European glaciers.

Svante Arrhenius, a Swedish physicist and chemist (1859–1927), was the first to link changes in carbon dioxide and surface temperature and the ice ages. Through ingenious scientific reasoning he deduced that halving the amount of carbon dioxide in the atmosphere would lead to a cooling of 4–5°C and a doubling of carbon dioxide would lead to a warming of 5–6°C. Considering Arrhenius published these results in the last throes of the 19th century it is remarkable that his values are anywhere near to present-day estimates calculated by state-of-the-science computer models that describe vertical motion and 3D processes. Many pioneering scientists later built on this early work.

As mentioned earlier, a wide range of other atmospheric constituents affect Earth's radiation budget. Aerosols are small particles or liquid droplets suspended in air or another gas, which can absorb or reflect solar radiation. As we discuss later (see 'Atmospheric water'), the interaction between aerosols and clouds can dramatically perturb the balance of incoming and outgoing radiation that can lead to cooling as well as warming Earth's surface.

Thermodynamics

Thermodynamics is the study of the relationships between heat and other forms of energy. It is used to help explain our quantitative understanding of the atmosphere from microphysical processes to global circulation patterns.

The atmosphere is a collection of gases that behave collectively and individually like an ideal gas—a gas that is composed of lots of particles moving rapidly in all directions that do not interact except when they collide. This allows us to describe the atmospheric state of a gas using its volume V, pressure P, and temperature T using the ideal gas law: $VP = nRT$, where n is the number of moles of the gas being studied and R is the gas constant.

The composition of the atmosphere is described here as a volume mixing ratio that is defined as the ratio of the partial volume of an individual gas to the volume of air. As stated in Chapter 1, most of Earth's atmosphere is nitrogen (78 per cent) and oxygen (21 per cent) with everything else residing in the remaining 1 per cent. For atmospheric composition we typically talk about parts per million (ppm, 1/1,000,000), billion (ppb, 1/1,000,000,000), or trillion (ppt, 1/1,000,000,000,000). The other metric of composition is partial pressure, which for an individual gas is the pressure exerted by the molecules of that gas if they alone occupied a volume V at temperature T.

The atmosphere never stops moving. Imagine a slab of air such that the horizontal scale of that slab is large compared to its vertical scale. For such a slab of air the gravitational force balances the buoyancy force, so that it is in hydrostatic equilibrium. This situation results in a mathematical relationship that describes how atmospheric pressure falls off exponentially with height, the hydrostatic equation, as a function of gravitational acceleration (g, 9.81 m/s^2) and the density of air.

Using the ideal gas law we can write an equation that involves temperature such that pressure decreases faster with altitude in colder atmospheres. A pressure scale height is defined as the height at which the pressure decreases by a factor of e (which has a value of approximately 2.718). In Earth's atmosphere the pressure scale height is approximately 8 km. If we suppose that atmospheric pressure on Earth's surface is 1000 hecto pascals, the atmospheric pressure would then be 368 hecto pascals at 8 km, and 135 hecto pascals at 16 km, and so on. For comparison, the approximate pressure scale height on Venus is 16 km, on Mars is 11 km, and on Jupiter is 27 km.

Two laws of thermodynamics are most relevant to studying the atmosphere. The first law describes the conservation of energy: energy cannot be created or destroyed but it can be transformed into other forms. The second law describes how an isolated system will move towards a state of greater thermal equilibrium. Most of what we understand about the atmosphere stems from these two laws.

To illustrate these principles of thermodynamics here and throughout the chapter, imagine a parcel of air. The air parcel has a certain pressure, volume, and temperature and is free to move. In addition to the kinetic and potential macroscopic energies of the parcel, we also have to consider the internal energy u of the parcel due to the potential and kinetic (mainly translational and rotational motions) energies of the molecules and atoms.

An increase in the internal kinetic energy, for instance, manifests itself as a macroscopic increase in temperature. After this parcel receives a certain amount of heat energy q it has the ability to do an amount of external work w. Work is defined as the ability of a force (here, pressure p) to move the object by a certain distance or more generally to change the volume V of that object. To conserve energy, any excess of energy over and above the external work done by the body must lead to an increase in the internal energy of the body. In other words: $\Delta u = \Delta q - p\Delta V$, where Δ is shorthand for a small change.

If our parcel of air can change its physical state (pressure, volume, or temperature) without a change in heat then we call that change adiabatic; if it occurs with a change in heat we call that change diabatic. Examples of diabatic processes include thermal conduction (but air is a poor conductor of heat), convection (rising warm air but geographically variable), advection (horizontal movement of air), radiation, and latent heat (due to phase changes of water).

Under adiabatic conditions, when air rises it expands because the pressure is lower at higher altitudes. As our air parcel expands it does work by pushing air that surrounds it. Because the air does work but receives no heat it must, following the first law of thermodynamics, lose internal energy. This exchange of energy translates into a decrease of temperature. Under these conditions temperature decreases with altitude at a rate of 9.8°C/km. This value is often called the dry adiabatic lapse rate, where by convention a positive lapse rate is defined as a decrease of temperature with altitude.

A related concept is the potential temperature. This is the temperature that our parcel of air would have if it expanded/compressed adiabatically from its existing pressure and temperature to standard pressure (generally accepted as 1000 hecto pascal). Potential temperature is a conserved quantity for adiabatic processes but changes with diabatic transformations, so it is a useful quantity for atmospheric thermodynamics.

Another way to thermodynamically describe our air parcel is to quantify its total heat content, or enthalpy H. Enthalpy includes the internal energy u of the parcel and the energy required to accommodate the parcel if the atmospheric pressure remains constant pV: $H = u + pV$. We can use that relationship to rewrite the first law of thermodynamics: $\Delta H = \Delta q + \Delta pV$. Enthalpy provides the central relationship for understanding atmospheric flow. At constant atmospheric pressure, heating the air parcel is equivalent to increasing its enthalpy that is equivalent to $\Delta u + p\Delta V$. The corresponding decrease in density of the air parcel results in wind.

The vertical movement of our air parcel is linked with how atmospheric temperature changes with altitude. To illustrate this point, think about a layer of atmosphere (Figure 8) with an environmental lapse rate that is steeper than the dry adiabatic lapse rate (denoted by G_d). Within that layer, as we move our parcel of unsaturated air from point O to a higher altitude point A, the temperature falls to T_A. The air parcel is then cooler than the surrounding environmental temperature. The air parcel will respond immediately to the reduction in atmospheric pressure with altitude. From the ideal gas law the displaced air parcel is denser than the surrounding air so the parcel will sink back to O.

Likewise, if we move our air parcel downwards its temperature will be warmer that the surrounding air and the parcel will rise back to O. The restoring force is proportional to the difference between the environmental lapse rate and G_d. An inversion layer, in which temperature increases with height, is an example of an extremely stable atmosphere. We will see later that inversion layers over cities impede the ventilation of air pollutants with implications for human health and visibility.

Conversely, for a situation where the environmental lapse rate is less steep than G_d (Figure 8) our parcel displaced upwards will

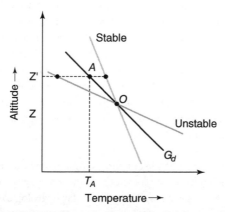

8. **Schematic illustrating the relationship between atmospheric lapse rates and static stability of air parcels.**

find itself warmer and less dense than its surroundings and continue to rise. Similarly, if the air is displaced downwards it will be denser than the surrounding atmosphere and continue to sink. This atmosphere is said to be statically unstable. These conditions are limited to the lowermost atmosphere. If the environmental and dry adiabatic lapse rates are equal the atmosphere is considered to be neutrally stable.

Atmospheric water

Water accounts for 0.25 per cent of Earth's atmosphere by mass—equivalent to a global puddle of 2.5 cm—but its ability to reflect incoming radiation and absorb outgoing longwave radiation makes it disproportionately important in the atmosphere. Water is the dominant atmospheric constituent responsible for the loss of radiative energy to space and hence atmospheric cooling.

Water is present in the atmosphere in three phases: gas, liquid, and ice (Figure 9). The energy released to the atmosphere when water changes from one phase to another, holding temperature constant, is called the latent heat of that phase change. Phase transitions are reversible. Latent heat is also released to the atmosphere when gas condenses to liquid and when liquid freezes to ice, and it is taken from the atmosphere to melt ice to liquid, to evaporate liquid to gas, and to sublimate ice to gas. Water has an unusually high latent heat due to the large amount of energy required to break hydrogen bonds in the condensed phase, and it is this that explains why water plays such a disproportionate role in Earth's atmosphere.

To illustrate the magnitude of energy contained by water, imagine that we could controllably translate Earth's 2.5 cm global puddle in latent heat energy averaged over a 10-day hydrological cycle. The result is an energy source that is many times greater than the annual energy use for the global economy.

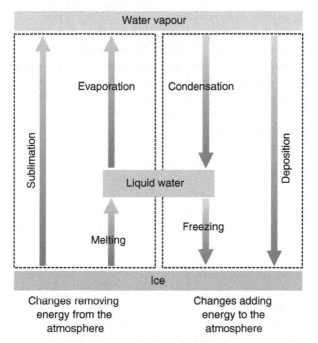

9. **The processes associated with transforming water into its different phases.**

The ability of water vapour (or any gas) to condense is described by its saturation vapour pressure. Imagine a sealed glass jar kept at a constant temperature. The jar contains a layer of water vapour sitting immediately above a layer of liquid water. Water molecules will condense into liquid onto the surface and liquid will evaporate from the surface into water vapour. This will continue until the number of water molecules condensing matches the number evaporating. At this equilibrium point the vapour above the liquid is saturated and the pressure of that vapour is called the saturation vapour pressure. All phases of water have a saturation vapour pressure. The degree to which water vapour is close to being saturated is measured commonly by its relative humidity.

The saturation vapour pressure increases approximately exponentially with temperature so that at warmer temperatures more water vapour can exist in the atmosphere without condensing. So that in a warming climate, the atmosphere will be able to hold more water and because water vapour is a strong greenhouse gas it can effectively warm Earth's surface that can result in more water being evaporated—this is an example of a positive climate feedback.

Water vapour alters the thermodynamics of a parcel of air. For example, we have seen that as air is elevated in altitude it cools by adiabatic expansion. If our air parcel now contains water, we find that as it cools the water begins to condense and release latent heat. Through buoyancy, this additional heat allows the parcel to rise higher than the dry air parcel.

Water vapour also affects the static stability of the atmosphere. To illustrate this, let's return to our air parcel. At the two extremes, the result for the moist air parcel is similar to the dry air parcel described earlier. If the environmental lapse rate is less than the saturated adiabatic lapse rate then the atmosphere is statically stable, and if the environmental lapse rate is larger than the dry adiabatic lapse rate then the atmosphere is statically unstable. The tricky situation is when the environment lapse rate lies between the saturated and dry adiabatic lapse rates in which the vertical stability depends on whether our air parcel is saturated.

Clouds are made of water droplets, ice, and/or particles, ranging in size from one micron to one centimetre, depending on cloud type and its origin and temperature. They form when the air is supersaturated due to cooling by air parcels ascending until they reach their dew-point temperature (quite literally the temperature where dew is formed) so that any additional water vapour condenses on aerosol particles. These particles act as condensation nuclei that grow as they collect more water vapour. The water cloud droplets then begin to grow through vapour diffusion.

As the droplets get bigger collision-coalescence processes are more important. Common aerosol particles include desert dust and sea salt, which are discussed in Chapter 4. There are a number of reasons that can lead to water-laden air rising, cooling, and forming a cloud. Solar heating causes air to rise. Air that is forced to rise over a mountain or hill has the same effect. Clouds can also form when a mass of warm air collides and rises up over a mass of cold, denser air. The boundary between these two masses is called a front. Weather over the UK, for instance, is influenced by westerly flow and associated fronts that bring bands of cloud and rain.

Our understanding of cloud and precipitation formation and development is far from complete. The overarching challenge lies in the fact that the development and evolution of clouds span an immense range of spatial and temporal scales so that explicit modelling of all cloud processes is difficult/impossible.

The feedbacks between clouds and climate, e.g. how the distribution of clouds will respond to a warming climate, represent the dominant uncertainty in climate models. Clouds directly affect Earth's atmospheric radiation budget. They scatter incoming shortwave solar radiation that acts to cool the planet and they absorb infrared radiation emitted from Earth's surface and lower atmosphere that eventually warms the planet.

The balance between these competing effects depends on the macrophysical (e.g. thickness and altitude) and microphysical (e.g. liquid droplets vs. ice crystals) properties of individual clouds. For example, clouds with more ice crystals will reflect more incoming solar radiation. High-altitude thick clouds are more effective at reflecting incoming solar radiation, but they also reduce the amount of infrared radiation lost to space from Earth's surface and lower atmosphere.

Climate models can capture the largest-scale observed variations in clouds but they are far from being able to reproduce observed properties and small-scale variations in clouds. This places significant limitations on our understanding of future climate.

Clouds also indirectly affect climate by interacting with aerosol particles (Figure 10). Our knowledge of these processes is arguably more uncertain than the direct effects. There are two main indirect effects of increased aerosols. From a phenomenological perspective the ideas are intuitive. In an unperturbed environment, where the number of aerosols is mainly due to natural processes, these cloud condensation nuclei lead to cloud formation, and the individual particles grow and eventually deposit within some form of precipitation.

In an environment where there is an additional source of aerosols (e.g. from industry) the same amount of water vapour in the atmosphere condenses on many more particles. This has two effects. First, the perturbed clouds comprise many more particles so that the cloud is effectively brighter and reflects more sunlight back to space; the extent to which depends on cloud depth. Second, the larger number of individual particles requires more time to grow to a size suitable for precipitation, which is associated with a slower particle collision frequency. Two processes are responsible for this. The primary reason is that smaller particles fall more slowly than larger particles, and the secondary reason is that smaller particles are less efficient at coalescence. This process also depends on the depth of the cloud structure.

An accessible example of this aerosol–cloud interaction is from aircraft condensation trails. The aircraft engine exhaust includes mostly water vapour and carbon dioxide, but also a small amount of particles that can promote the growth of water droplets. The combination of these water droplets and the low atmospheric temperatures at cruise altitude can lead to the formation of ice particles that are visible as white streaks left

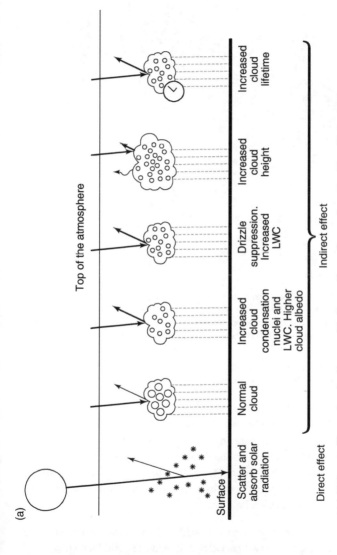

10. (a) Diagram illustrating the direct and indirect effects of aerosols on Earth's radiation budget.

(b)

(b) **Image of cloud streaks over the north-east Pacific Ocean, 4 March 2009, formed by aerosol particles from ship exhaust. The image is based on observations collected by the Moderate Resolution Imaging Spectrometer (MODIS) on NASA's Terra satellite.**

behind by aircraft. Despite their streakiness they significantly affect Earth's radiation balance.

Ship exhaust emissions, which remain unregulated, also exhibit trails that can be observed from space. Part of their exhaust includes material that can form the basis of cloud condensation nuclei. The net result is the formation of very bright clouds that follow the track of ships (Figure 10). As the commercial shipping fleet continues to grow so does the global impact of ship emissions on Earth's radiation balance.

The semi-direct effect is associated with the aerosol composition. It is possible the aerosol particles that seed the cloud formation

42

absorb radiation. So once the cloud begins to grow, air containing the water droplets can heat up, evaporate, and shrink the cloud. The importance of the semi-direct effect is also sensitive to whether the absorbing aerosols represent individual constituents that have combined together chemically and/or physically or whether they coexist within the cloud. Current research is focused on reducing the uncertainties associated with aerosol–cloud interactions.

Atmospheric electricity

Fair weather (clear-sky) electricity describes Earth's background electric field that exists in the lower atmosphere with an approximate value of 100V/m, falling off exponentially to about 1V/m in the stratosphere. Generally, the atmosphere is positively charged and the surface is negatively charged. Earth's atmosphere stores about 150×10^9 J of electrical energy.

All clouds contain some level of electric charge over and above the fair weather electricity value. Precipitation particles can induce an initial amount of charge from the fair weather electric field with equal amounts of positive and negative charge. The dominant process responsible for charge separation within a cloud is the collision of ice particles that exist in a variety of forms.

Graupel forms when a falling ice crystal collides with super-cooled cloud droplets. The result is an ice particle covered by frozen cloud droplets, forming an almost spherical particle that is like a hailstone but much less dense. Graupel collides with ice crystals in the presence of super-cooled cloud droplets. The graupel particles end up with a net positive or negative charge depending on the environmental temperature. At temperatures lower than –15°C graupel particles become negatively charged and the ice crystal becomes positively charged; at higher temperatures the polarity of graupel particles and ice crystals is reversed.

Larger graupel particles fall faster than ice crystals that have not collected any super-cooled water droplets, and so they tend to

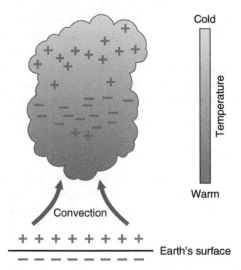

Cold

Temperature

Warm

Convection

+ + + + + + +
———————————————— Earth's surface
– – – – – – – –

11. **A schematic of the tripole charge separation structure in thunderstorm clouds due to collisions between graupel particles and snow crystals.**

separate after they collide. The negatively charged graupel particles accumulate in the middle of the cloud and the lighter snow crystals are lofted higher into the cloud due to the updraught of the convective cloud. In practice, other processes also play a role in distributing charge within a cloud.

The net effect is a tripole charge separation within the cloud with a net positive charge at the top, a net negative charge in the middle, and a small amount of positive charge towards the warmer bottom of the clouds where the graupel–ice crystal charge separation is reversed (Figure 11). Most thunderstorm clouds have this broad-scale tripole feature, particularly the negative charge centre around –15°C, independent of variations in cloud base height or cloud height. More complicated structures are often observed.

Clouds need to be 3–4 km in vertical thickness before we observe significant electrification. Air acts as an insulator for the charge difference in the cloud and between the cloud and the Earth's surface. The high density of negative charge at the cloud base repels the electrons near the Earth's surface, leaving a positive charge. The associated electric field induced by this charge difference grows in strength. At a certain threshold (~3 MV/m, which is the dielectric strength of air) the insulating properties of air break down dramatically and it becomes for a millionth of a second a conductor.

While lightning is familiar to all of us, the initiation of the next steps is not fully understood. Two channels of ionized air (called stepped leaders) develop from within the cloud when the charge difference becomes large between the main region of negative charge in the middle of the cloud and the weaker positive charge in the cloud base. At this point the insulating capacity of air breaks down and the negative stepped leader moves towards the Earth.

The negative stepped leader moves in fits and starts (at a speed of about 120 km/s) with each successive spurt, concentrating the charged particles at the very ends that often split into new leaders in a tree-like pattern. The zigzag movement of these stepped leaders speeds up, as they get closer to the surface.

As the negative stepped leader approaches the ground the strength of the electric field increases with the presence of the opposite charge at the surface. If the field is strong enough a positive upward streamer can be formed. Once the downward leader meets an upward streamer the lightning channel is complete and electrical charge can flow rapidly. What follows is a massive discharge of electricity and bright flash of light. The negative charge accumulated in the stepped leader is rapidly transferred to the ground. The lightning flash propagates from the ground to the cloud. This is called the return stroke. Any residual negative charge left in the cloud where the return stroke meets the clouds can be rapidly transported to the ground via established channels.

The massive flow of charge, and the speed at which it travels, results in the air being superheated momentarily to temperatures hotter than the Sun. This causes the charged particles to glow blue-white (lightning) and the air to be superheated and to expand explosively leading to a sonic shock wave (thunder). Because of the difference in the speed of light and sound, lightning always precedes thunder. If you count the number of seconds (one hippopotamus, two hippopotamus,...) between seeing the lightning and hearing the thunder and divide by three you get the distance away you are from the storm; this works because the speed of sound travels at roughly a third of a kilometre every second.

The vast majority of lighting discharges occurs within clouds, with only 20 per cent associated with cloud to ground. Lightning strikes occur somewhere on Earth more than forty times per second, equating to more than a billion strikes per year. They form preferentially over geographical regions with intensive heating where it is typically hot and humid, i.e. over landmasses and over the tropics. As we shall see in Chapter 4 they also play a role in chemistry because lightning can transfer sufficient heat to pull apart bonds of stable molecules such as nitrogen.

Charged particles represent a very small fraction of the atmosphere but they play a disproportionate role in determining its behaviour. The study of atmospheric electricity has a long and illustrious history with documented experiments starting in the 1750s with polymaths Benjamin Franklin in Philadelphia and Thomas-François D'Alibard in Paris. The study of thunderstorms has helped to drive our understanding of atmospheric electricity but thunderstorms are only a part of the global atmospheric electric circuit, which describes the continuous flow of electric charge driven by the voltage difference between the ionosphere and the Earth's surface. Thunderstorms are transient (dramatic) current generators that cause an accumulation of positive charge in the ionosphere.

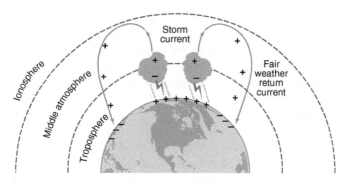

12. Simplified schematic of the global electric circuit incorporating the transient storm current and the fair weather return current.

Later, in the 1770s, measurements showed the existence of positive electrification on fine days. This fair weather atmospheric electricity has been studied ever since and describes the slow return of the positive charge back to the surface in non-thunderstorm conditions. Extensive ship-borne measurements that produced the Carnegie curve, show that fair weather atmospheric electricity has a clear diurnal cycle peaking at 19.00 Universal Time (UT) with a minimum at 0300 UT. The periodicity in the cycle is thought to be due to regular variations in atmospheric electrification due to different weather regimes around the world. Subtle changes in these variations have been attributed to large-scale climate variations such as the El Niño–Southern Oscillation (ENSO).

There is still much to be understood about the global electric circuit (Figure 12). Recent studies have investigated the importance of transient luminescent events. The Scottish physicist C. T. R. Wilson in the 1920s first predicted the electrical breakdown of the atmosphere high above large thunderstorms and sightings have been reported since then, but it was not corroborated with direct measurements until 1989 with the advent of low light-level television equipment. These so-called sprites are triggered by the discharge of positive lightning at the top of the thunderstorm clouds. They appear as luminous pink-red flashes and occur in clusters at 50–90 km above the Earth's surface high above the thunderstorm cloud. They are classified by their appearance: jellyfish sprite with a bell and downward tentacles, carrot sprites, and column sprites.

Blue jets are similar but protrude from the top of the thunderstorm cloud in a narrow column to the base of the ionosphere. ELVES (an acronym for Emission of Light and Very low frequency perturbations due to Electromagnetic pulse Sources) are a flattened and stretched doughnut (400 km across). A space shuttle mission first reported them in 1990 above a thunderstorm off the coast of French Guyana.

The lightning discharge in the cavity, defined by the volume of air residing between Earth's surface and the ionosphere, can resonate in Earth's electromagnetic field at extremely low frequencies. Much like we use the ionosphere as a method of transmitting signal around the world, monitoring these frequencies provides information about the frequency of lightning on a global scale. However, pinpointing the location of individual lightning discharges is difficult.

Chapter 3
Atmospheric motion

Solar activity is the main driver for Earth's large-scale atmospheric motion. Similarly, stellar activity is the driver of planetary atmospheric motion. The annual averaged incoming solar radiation per unit area is greater at lower latitudes than at higher altitudes due mainly to the Earth's tilt that ranges from 22.1° to 24.5°. Lower latitudes receive more energy from the Sun that they emit back to space, while the higher latitudes emit more radiation back to space than they receive directly from the Sun. On a global scale the Earth is in approximate thermal equilibrium, which implies that energy is being transported from low to high latitudes.

The thermal gradient between the tropics and the poles drives the hemispheric circulation. Based on the discussion in Chapter 2 about atmospheric thermodynamics, a tropical air parcel absorbs energy from thermal radiation, increasing its internal energy and/or increasing its volume, with a corresponding decrease in density. This allows the parcel to rise in the atmosphere to a level where it is statically stable. The warm air that has risen from the surface moves horizontally to colder, higher latitudes driven by the pressure force, reducing the surface pressure at the location where the air has risen.

Similarly, an air parcel that travels to higher latitudes becomes cooler, contracts, and therefore becomes denser than its surrounding

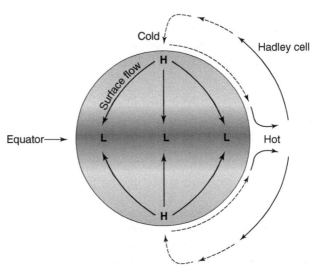

13. General circulation of Earth-like planetary atmosphere, in the absence of rotation, driven by differential solar heating. Low-and high-pressure regions are denoted by L and H, respectively.

environment, and consequently sinks in the atmosphere resulting in high pressure at the location where it sinks. The surface pressure gradient between high and low latitudes results in a net force that is directed from high to low pressure so that some of the descending air travels equatorward thereby closing the circulation cell. In the absence of rotation and assuming a uniform distribution of land surface, the large-scale movement (general circulation) of the atmosphere comprises two hemispheric-scale circulation cells (Figure 13) generated by the differential solar heating due to the Earth's tilt.

The role of rotation

But Earth is rotating and is composed of land and ocean. Further, land surfaces, including mountains, modify these large-scale circulation cells. Earth orbits around the Sun and rotates about its axis of rotation that goes through the North and South poles (distinct from the magnetic poles). Earth rotates from west to east and completes a full rotation approximately once every 24 hours with respect to the Sun. Earth's rotation is slowing down due to the pull of the Moon, but only by about 1.7 millionth of a second every century.

On a rotating Earth, air moves from high surface pressure to low surface pressure, as before, but it is deflected by the rotation of the planet. Imagine you are on a child's spinning roundabout and you have a friend standing at a fixed point away from the roundabout (Figure 14). Your friend throws a ball towards the centre point. At the same time the roundabout is spinning anticlockwise so that the ball veers off on a straight path away from the centre. This is clear to see from the fixed observer frame. From your perspective as the rider on the rotating roundabout the ball will appear to deflect from a straight path. From your point of view what happens appears to require a fictitious force. The Coriolis force is that fictitious force, which is necessary to describe Newton's second law in a rotating reference frame (on Earth this is where we stand).

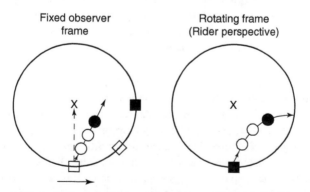

14. **An illustration of the Coriolis force. The path of a rolling ball (circle) on a playground roundabout as observed in (left) a fixed observer frame and (right) the rotating frame. The rider is denoted by a square. Open symbols denote past positions of the ball and rider.**

The same principle applies to a parcel of air. Look at the schematic again, but now imagine you are looking down at the North Pole. The rotation of the Earth from west to east deflects air in a clockwise direction. If you were looking down at the South Pole Earth's spin would deflect air in an anticlockwise direction. Because Earth rotates slowly the force is small and only perceptible for large-scale air motion.

From a physical perspective, the ball in our example is deflected to conserve its angular momentum, which is the rotational analogue of linear momentum (given by the product of mass and velocity of an object). Angular momentum describes the mass and velocity of a rotating object but also includes the relative position of that object from the axis of rotation.

Let's return to our parcel of air, this time travelling around the equator. Assuming that the Earth is a sphere, the air parcel is at the furthest distance away from the axis of rotation equal to the Earth's radius R. As the air parcel moves to higher latitudes by the pressure force it gets closer to the axis of rotation by a factor of the cosine of the latitude. For example at 60°N the air mass is 50 per cent closer to the axis of rotation. At higher latitudes the air mass conserves angular momentum by increasing the eastward component of its velocity relative to Earth.

From an observer perspective, this would appear as if the air parcel was deflected to the right. As the Coriolis force and pressure gradient force begin to balance, the air parcel starts to move at a constant velocity. This equilibrium between the two forces is called geostrophic balance and leads to a geostrophic flow that is parallel to lines of constant pressure (isobars). In the northern hemisphere air in geostrophic balance travels clockwise around a high-pressure centre (also known as an anticyclone or high, Figure 15), and anticlockwise around a low-pressure centre (also known as a cyclone or low).

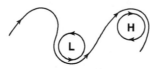

15. Schematic of geostrophic airflow in the northern hemisphere and the position of high-pressure (H) anticyclonic weather systems and low-pressure (L) cyclonic weather systems.

Near the surface, where the atmosphere is in contact with the surface, there is a friction force due to local terrain (e.g. mountains, buildings, ocean waves), which acts in the direction opposite to the airflow and tends to slow down the geostrophic flow. The effect of this frictional deceleration leads to air being deflected into low-pressure cyclones and deflected out of high-pressure anticyclones. Cold air from above sinks to fill the space left for the air diverging from a surface high. The spiral of descending air, warmed by compression according to the first law of thermodynamics, results in light winds and reduces the formation of cloud; calm weather conditions.

In contrast, air converging onto a surface low requires air to rise (much like squeezing the bottom of a tube of toothpaste). Conversely, as the air rises it cools and expands, and water condenses to form clouds and potentially rain (Figure 16). The movement of air will behave differently depending on whether it is strongly stratified by density.

Earth's rotation modifies the general circulation so that it can be described conveniently as three circulation cells: the Hadley cell, the Ferrel cell, and the Polar cell (Figure 17). Solar heating drives the upward branch of the overturning cell over the tropics. It is marked on Earth's atmosphere as a narrow channel of clouds (a few 100s km wide) where there is persistent convergence of air and thus ascending motion. This is called the intertropical convergence zone (ITCZ) and it meanders on a seasonal and year-to-year basis. Tropical rainforests are typically found

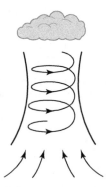

16. Schematic of a cyclonic low-pressure system.

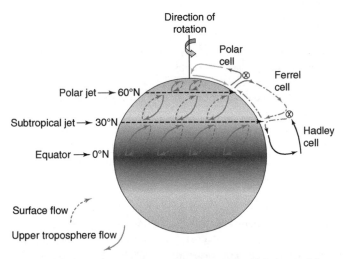

17. General circulation of Earth's atmosphere. The circle inset with an X denotes that the direction of flow is into the page. The Ferrel cell is denoted by dashed lines as a reminder that it depends on the Hadley and Polar cells. Darker colours denote warmer temperatures.

underneath the ITCZ because of the precipitation associated with the rising warm air.

The poleward branch of the cell extends to about 30 degrees latitude before descending. By conservation of angular momentum the air parcels accelerate as they move northward until they reach a maximum at midlatitudes where they break down into large eddies. This tropical overturning cell is called the Hadley cell, named after George Hadley who was an English lawyer and an amateur meteorologist during the 18th century. Hadley originally described a much larger cell that stretched further to the poles, but he did not take into account the conservation of angular momentum. Weather below the descending branch of the Hadley cell is characterized by little precipitation and calm winds. This is where the majority of deserts are located.

Variations in the latitudinal extent of the Hadley cell have been linked with changes in climate with potentially dire consequences for vegetation, agriculture, and habitability. Near the surface, the Coriolis force deflects equatorward air westward resulting in the trade winds, which have been exploited for centuries by explorers and traders. In the upper troposphere, the poleward transport of air in the Hadley cell turns eastward by the Coriolis force, eventually mixing with cooler midlatitude air.

When the atmosphere is in geostrophic balance and hydrostatic balance, the negative meridional atmospheric temperature gradient (along a line of constant longitude) is related to a positive vertical wind shear with altitude. This thermal wind relationship helps to explain the emergence of the jet streams, which are fast and narrow air currents that are close to the troposphere where there are large horizontal temperature gradients due to the mixing of large-scale circulation cells. The speed of the jet stream varies with the horizontal temperature but is typically greater than 100 km/h. The mixing of tropical and midlatitude air in the descending branch of the Hadley cell results in the westerly subtropical jet stream.

The Polar cell is the northernmost of the three circulation cells. It typically stretches from 60 degrees to the pole and behaves much like the Hadley cell. At the pole, cooler high-altitude air sinks causing high surface pressure, eventually returning southward via the low-altitude branch of the cell, and a compensating rising branch at relatively warm midlatitudes. The Coriolis force is strongest at the poles so the southward air deflects to the right, which forms the surface polar easterlies. There is typically a much larger horizontal temperature gradient associated with the mixing of tropical (transported from the midlatitudes) and polar air, resulting in a much stronger polar jet stream. The jet stream can stretch for thousands of kilometres and can be hundreds of kilometres wide. It is this rapid air current that is often exploited by commercial air traffic to hasten transatlantic flight time.

Planetary-scale horizontal meandering of the polar jet stream, associated with warm air travelling poleward and cool air travelling equatorward, is called an atmospheric Rossby wave. These waves were named after the Swedish meteorologist Carl-Gustaf Rossby (1898–1957) who mathematically explained their existence in the atmosphere and oceans. Such waves are responsible for the exchange of information between widely separated geographical regions, so-called teleconnections. The horizontal distance between a peak and trough (half a wavelength) of a stationary Rossby wave is about the width of continental North America (~2,500 km). As a consequence when there is anomalous poleward transport of air over the west coast of North America there is typically equatorward transport of air over the east coast of North America, which we will see later.

The middle Ferrel cell is named after William Ferrel, an American meteorologist who in 1856 published his theory of midlatitude dynamics in a medical journal. It is a thermally indirect circulation cell and behaves differently to the other cells: it is characterized by sinking air at relatively warm midlatitudes and rising air at relatively cool high latitudes. In other words, the existence and strength of the Ferrel cell circulation depends on the motion of the Hadley and Polar cells. The Ferrel circulation is composed by eddies, deviations to the smooth, mean flow of the atmosphere.

Travelling weather systems, such as cyclones and anticyclones, are generated by the descending and ascending branches of the Hadley and Polar cells. These eddies play an important role in transporting heat and momentum from the tropics to midlatitudes and maintaining the global circulation. Near the surface the Coriolis force deflects the poleward air eastward resulting in the westerlies, which can be thought of as being maintained by a series of large low-pressure centres that spawn smaller cyclonic eddies. Low-pressure cyclones at high latitudes transport warm moist air poleward that can result in precipitation. High-pressure anticyclones represent cool air being transported to lower latitudes.

Both jet streams meander meridionally and sometimes the flow can generate a stationary Rossby wave pattern such that pressure centres persist over a region for several days, reinforcing the existing conditions, and this can lead to extreme weather. For example, the persistent configuration of the polar jet stream over North America during winter 2015 led to a situation where Arctic air was pulled down over New England and the low-pressure centre spawned cyclonic weather systems that produced a larger number of snowstorms. While the western United States experienced anomalously warm weather. See Box 1 for some approximate timescales associated with atmospheric transport.

Role of oceans

So far, we have considered a rotating Earth that has uniform pressure at a particular latitude. There are of course also pressure gradients in the zonal directions (along a line of constant latitude) caused by land-surface ocean heterogeneity. Chances are that

Box 1 Transport timescales in the atmosphere

Within the mean general circulation of the atmosphere there is a range of characteristic timescales and corresponding spatial scales associated with typical phenomena (e.g. fronts, hurricanes). First, it takes about two weeks to circumnavigate the globe at midlatitudes, and on average it will be shorter at higher latitudes (shorter distance) and longer at lower latitudes (longer distance). This timescale is relevant to the transport of pollutants emitted from one country and received by another. Transporting air from midlatitudes to either polar or equatorial latitudes takes about one or two months. For air to cross from the northern hemisphere to the southern hemisphere takes about a year on average but occasionally there is rapid movement of air between the hemispheres. In this case, the ITCZ effectively acts as a barrier. Because the ITCZ meanders as far as 500 km from the equator it is possible that northern hemisphere (generally more polluted) air can be temporarily located in the geographical southern hemisphere and vice versa.

you have probably experienced the circulation resulting from these land–water pressure gradients before, at least in its gentlest form as land–sea breeze (Figure 18), which is a wind between water bodies and nearby land (e.g., seaside or lakeside).

A sea breeze is caused by a gradient in air pressure between the water and land determined by their different heat capacities (the amount of heat required to raise the temperature of an object by one degree Celsius). Water has a higher heat capacity than land due to the energy required to break the hydrogen bonds in water molecules. It means that a certain amount of water takes longer than land to heat up but it also means that the water cools down slower than land. As a result, in the morning when the Sun rises the land will heat up quicker than the water because of its lower heat capacity. The atmosphere overlying

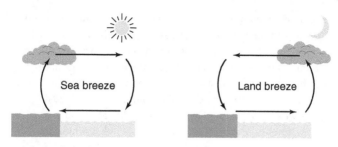

18. The land–sea breeze circulation pattern.

the land will begin to warm via conduction and increase atmospheric pressure at higher altitudes (e.g. 100 m). Rising, moist air will help form clouds. The pressure gradient at 100 m generates a movement of air from the land to the water. This movement of air decreases the surface pressure over the land and increases it over the water, resulting in a counter-movement of air at the surface from the water to the land. To replace the air displaced over the water, the air above sinks to complete the circulation. For the same reasons, during the night the land loses heat more quickly than the ocean so a reversed circulation is established.

Monsoons are essentially large-scale land–sea breezes where the associated land–ocean surface temperature gradients are much larger. These gradients are sustained because the surface ocean is responding to changes in solar radiation and also because the surface ocean is mixing with deeper waters which occurs on a timescale much longer than a day. The major monsoon systems are over West Africa and Asia-Australia. The West African monsoon is driven by seasonal shifts in the ITCZ and the temperature gradients between the Sahara and the equatorial Atlantic Ocean. The Asian-Australian monsoon is composed of interacting monsoon systems stretching from Australia to Africa and from northern Australia to the Russian Pacific coast.

The largest system is the Indian monsoon. As the land heats up during the summer months, moist air from the Indian Ocean is drawn towards the Himalayas. Convergence of warm air against the Himalayan mountain range forces the incoming air to rise, cool, and release the moisture as rain. Monsoon systems typically represent a substantial fraction of annual rain over affected regions with implications for their agricultural productivity. The vast amount of water that accumulates in a few months during the rainy season can overcome flood defences, and can trigger landslides.

The Southern Oscillation, the atmospheric component of the atmosphere–ocean coupled El Niño–Southern Oscillation (ENSO), is described by an oscillating surface pressure gradient between the eastern and western tropical Pacific Ocean (typically, every 2–7 years). This is not driven by changes in land and ocean temperatures but changes in temperature and winds over the ocean. Low atmospheric pressure tends to occur over warm waters, driven in part by deep convective cells, and higher atmospheric pressure over cold waters. Under normal conditions sea surface temperatures are warmer over the western Pacific. This results in a circulation, driven by the pressure gradient force, with surface easterly winds, rising warm air over the western Pacific, and descending air over the eastern Pacific. Other tropical zonal overturning circulations link all major continents and oceans. UK meteorologist Gilbert Walker, while based in India in the early 20th century as the Director of Observatories, inferred this (Walker) circulation using sparse surface observations. Changes in the Walker circulation reflect changes in the gradient in sea surface temperature (Figure 19). The mechanisms that drive these changes in sea surface temperatures are an active area of scientific research.

There are two phases of ENSO. One is the cold phase, called La Niña (meaning little girl in Spanish), and the other is the more familiar warm phase, called El Niño (meaning little boy). There

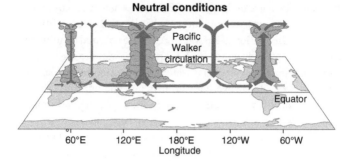

Neutral conditions

Pacific
Walker
circulation

Equator

60°E   120°E   180°E   120°W   60°W
Longitude

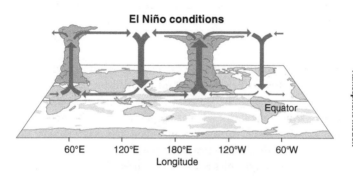

El Niño conditions

Equator

60°E   120°E   180°E   120°W   60°W
Longitude

Atmospheric motion

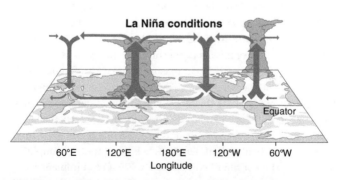

La Niña conditions

Equator

60°E   120°E   180°E   120°W   60°W
Longitude

19. Schematic of the coupling between ENSO and the Walker circulation.

are many downstream impacts of ENSO but here we focus on the immediate atmospheric phenomena. During La Niña conditions, the Walker circulation is enhanced relative to normal conditions, the stronger easterly winds push more warm water to the west, promoting upwelling of cooler water over the eastern Pacific. This cooler water cools the overlying atmosphere and impedes the formation of cloud and rain. Over the western Pacific, the air is warmed by the warmer waters increasing the chance of rain. During El Niño conditions, the Walker circulation is weakened relative to normal conditions or in extreme cases reverses. Weaker easterly winds are accompanied by elevated warming of the central and eastern tropical Pacific Ocean. There is still much we do not understand about ENSO such that our ability to predict the onset any one phase is limited.

The Madden–Julian Oscillation (MJO), discovered in 1971 by Americans Roland Madden and Paul Julian, is the largest source of atmospheric variability on intra-seasonal timescales. It involves a large-scale coupling between atmospheric circulation and deep tropical convection. The oscillation is the result of interactions between individual convective systems with a size of 100–1000 km with large-scale dynamics at the scale of ocean basins. The MJO modulates tropical precipitation and indeed it can be characterized by alternating bands of elevated and suppressed rainfall, mainly over the Indian and Pacific Oceans. Unlike ENSO, the MJO propagates eastwards over the warmest parts of these Oceans, at a speed of about five metres per second eventually crossing the tropics in 30–60 days. For that reason, the MJO is often referred to as the 30–60 day oscillation. The oscillation can send ripples through the atmosphere that are related to flooding, snowstorms, and hurricanes. Much like ENSO, the underlying mechanisms that initiate and propagate the MJO and its manifold effects on climate are still not well understood. See Box 2 for an example of how the MJO or the El Niño phase of ENSO can influence subsequent weather patterns.

Box 2 Atmospheric rivers and air pollution karma

Atmospheric rivers describe narrow filaments (few hundreds of kilometres wide and thousands of kilometres long) of water vapour travelling in the lower 2 km of the atmosphere at midlatitudes. There are typically a few of these rivers in a hemisphere at any one time, and they are responsible for the vast majority of water vapour transport outside of the tropics. In some instances these rivers contain as much water as the Amazon river, but most are much smaller. While they represent an integral part of the global water cycle, providing rain and snow, the larger atmospheric rivers are associated with the most extreme rainfall and flooding. The most (in)famous atmospheric river is called the Pineapple Express, owing to its apparent origin over the central Pacific near the Hawaiian Islands, which can seriously affect weather over the Pacific Northwest and California. Conditions that precede the Pineapple Express involve (1) elevated tropical deep convection over the western Pacific, either from the MJO or during the El Niño phase of ENSO, and (2) a jet stream configuration that encourages transport of water to the Northeast Pacific. A high-pressure system over the Gulf of Alaska blocks this polar jet but during the winter it weakens and shifts westwards. This splits the jet stream and increases the westerlies on the southern edge of the jet. The resulting midlatitude trough off the western coast of North America connects with the moisture plumes from the tropics bringing huge rainfalls that can result in flooding and landslides. There remains some debate in the scientific community about how these filaments form in the atmosphere.

The atmospheric general circulation also transports trace gases and particles lofted into the atmosphere from natural and human sources. The atmospheric lifetimes of these constituents can be long enough to travel thousands of kilometres so that pollution

(Cont.)

Box 2 Continued

emitted from one country can impact countries downwind. It only takes a few weeks to circumnavigate the globe so that all countries receive air pollutants from everyone else, with concentration reduced by atmospheric mixing. We can now see this long-range transport of pollution from space-based observations. As stricter air quality regulatory standards are imposed by national governments the difference between meeting and exceeding these standards can be determined by the pollution that has been emitted far upwind. The more difficult situations occur when the culprit is a natural source that is determined by geological (e.g. volcanoes) or climate (e.g. forest fires) processes. The volcano Eyjafjallajökull erupted in 2010 with little warning and because we had only a limited ability to observe particles and gases that could potentially interfere with jet turbines without intercepting the plume itself the European air sector was closed to all aircraft, and estimated to cost hundreds of millions of pounds per day. Even small changes in the location or timing of emissions and/or atmospheric transport can significantly influence air quality elsewhere. For instance, in 2006 due to a late snow melt in eastern Europe agricultural burning was delayed by a few weeks. This delay in burning coincided with a jet stream that picked up those emissions and rapidly transported them to higher latitudes, which resulted in record-high levels of air pollution over Scandinavia.

Chapter 4
Atmospheric composition

Nitrogen, oxygen, and argon collectively represent more than 99.9 per cent of the air we breathe. The gases relevant to climate and human health that sometimes dominate the headlines are all described in that remaining 0.1 per cent of air. But Earth's atmosphere hasn't always had that composition—it is on at least its third distinctive atmosphere.

A brief history of Earth's atmosphere

Earth received its first atmosphere about four and a half billion years ago when it was very new. This atmosphere was composed almost exclusively of hydrogen and helium, reflecting the composition of material around the Sun where Earth was formed. The atmosphere also included all the material we use as the building blocks of life (e.g. carbon, iron, calcium), but these gases were too light and fast moving for Earth to retain so they eventually drifted off in to space (see Chapter 1).

Earth started to generate its own second atmosphere less than four billion years ago. Early Earth was an extremely violent place, with much more volcanic activity than we see today. Volcanoes release water, carbon dioxide, and a range of organic gases. The atmosphere sustained very little oxygen with the possible exception of small niche microbial environments that supported oxygen production. Carbon dioxide dissolved in the oceans that formed early in Earth's history.

The third atmosphere is our current one that developed rapidly over a geological timescale of 600 million years or so. The first jump in oxygen started around about 2.45 billion years ago from essentially nothing to between 1–40 per cent of our present-day oxygen levels. This dramatic change was followed by a billion years or so when atmospheric oxygen levels did not change much, although much was afoot in the biological world.

Finally, just over half a billion years ago oxygen levels reached present-day levels. For this last change we have more data to inform our understanding of what happened. There is evidence that oxygen levels varied between 60 per cent and 160 per cent of present-day levels. This does not look like much of a constraint, but consider that combustion cannot be sustained at 60 per cent of present-day oxygen values and fires are not easily quenched at 160 per cent, so the development of forest ecosystems, for instance, would be difficult.

A simple box model of atmospheric composition

When we observe variations in gases and particles in the atmosphere they reflect changes in emissions and deposition, atmospheric chemistry (production and loss), and atmospheric motion. Let's now imagine a global atmospheric box that sits directly above Earth's surface (Figure 20). The concentration of a chemical X within that box at a particular time will be determined by emissions of X, deposition of X, and finally the chemical production and loss of X. To make things simpler we'll assume that the box is the height of the atmosphere so we can ignore vertical motion.

From this simple model, we can illustrate two important concepts in atmospheric composition that we will use later: (1) atmospheric lifetime τ and (2) mass balance. The atmospheric lifetime of a

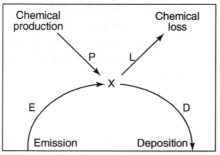

Global atmospheric "box"

20. **The concentration of an atmospheric chemical X in a global atmospheric box is determined by emission, deposition, and the chemical production and destruction of X.**

pollutant is important to know because it will have consequences for the impact of that pollutant on the physical, chemical, or biological components of the Earth system. Similarly, the concept of mass balance is important to understand because any imbalance will result in the atmospheric increase or decrease of that pollutant.

The lifetime of a gas X is the average time it remains in the atmosphere. Processes that remove X in our model, and therefore determine the lifetime, include deposition and chemical loss. The atmospheric lifetime of X is calculated by dividing the mass of X within the box by the removal rate (mass per unit time). If for example there are two competing removal processes the resulting value for τ is given by $1/\tau = 1/\tau_1 + 1/\tau_2$.

The lifetimes of gases range from seconds for the hydroxyl radical (OH), through weeks for tropospheric ozone and carbon monoxide, to decades for chlorofluorocarbons and greenhouses gases. These lifetimes have implications for the distance that the gases can travel across the globe.

The mass of X within the box is determined by its production and loss terms. If the production P of X equals the loss L of X ($P = L$)

within the box then the amount of X remaining in the box will remain the same, i.e. whatever was removed is replaced. This situation is called mass balance and describes a steady-state situation. If $P > L$ then the mass of X will have a positive atmospheric growth rate, and if $P < L$ the mass of X will decline with time, i.e. have a negative growth rate.

Stratosphere

The stratosphere is a region of the atmosphere that we rely on for our survival. The stratospheric ozone layer shields harmful ultraviolet-B light (spanning wavelength range 280–315 nm) penetrating to the surface, thereby protecting humans and ecosystems from harmful ultraviolet radiation. Without this layer life on Earth would possibly be on a very different evolutionary path.

Ozone comes from the Greek word *ozein*, 'to smell'. It was the smell that first alerted the German chemist Christian Schönbein to his discovery of ozone in the mid-19th century. Observations of the solar spectrum taken from Earth's surface first alerted scientists to the absorption of sunlight by gases in the atmosphere. Based on these observations, Walter Hartley hypothesized the presence of ozone in the atmosphere later that century.

But it was not until 1921 when new observations were reported that appeared to support Hartley's hypothesis about atmospheric ozone. Gordon Dobson and Frederick Lindeman at Oxford found that, while studying meteor trails, temperature increased with altitude above the troposphere, contrary to then current beliefs. Dobson correctly surmised that the source of heating was from the absorption of ultraviolet solar radiation by ozone.

What followed is an amazing scientific journey that resulted in the 1995 Nobel Prize in Chemistry for better understanding of the formation and decomposition of ozone, and in the eventual global stewardship of stratospheric ozone.

After Dobson and Lindeman's measurements, the next natural question was: why is there ozone in the stratosphere in the first place? In 1930, the widely accomplished scientist Sydney Chapman put forward a simple chemical mechanism comprising four reactions that describes the underlying reason for atmospheric ozone:

$$(1) \quad O_2 + h\nu \rightarrow O + O \, (\lambda < 242\text{nm})$$

$$(2) \quad O + O_2 + M \rightarrow O_3 + M$$

$$(3) \quad O_3 + h\nu \rightarrow O_2 + O \, (\lambda < 320\text{nm})$$

$$(4) \quad O_3 + O \rightarrow O_2 + O_2$$

Molecular oxygen (O_2) can be disassociated by photolysis by a high-energy photon (denoted by $h\nu$). The resulting atomic oxygen atoms (O) then react with O_2 to produce ozone (O_3). M is a third body that stabilizes the excited reaction product, and in the atmosphere this is almost always O_2 or nitrogen (N_2). In the Chapman mechanism there are two losses for O_3: one is photolysis that returns O_2 and O, and the other is the reaction with O that returns two O_2 molecules. Ozone bonds are relatively weak so less-energetic photons can break them apart. Reactions 2 and 3 occur very quickly (so much so we assume them to be in steady state) so we typically sum O_3 and O together to form odd oxygen O_x, i.e. O_3 + O. Production and loss of ozone is determined by the slower reactions 1 and 4.

The four reactions that constitute the Chapman mechanism help to explain the concentration of ozone in the upper stratosphere. The theory qualitatively describes the vertical distribution of ozone but the model incorrectly predicts the magnitude of stratospheric ozone. In the lower stratosphere we cannot assume O_x is in steady state and here atmospheric dynamics plays a role in stirring up chemicals. In the upper stratosphere, the theory overestimated ozone by a factor of two. Recognition of these weaknesses of the Chapman mechanism has since led to great scientific advances.

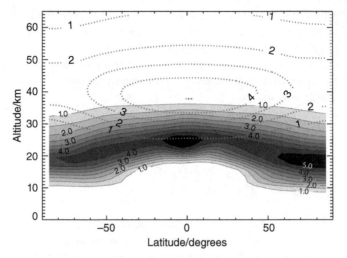

21. **Annual means of ozone concentration (10^{12} molecules/cm³, greyscale shaded contours) and odd oxygen production rate (10^6 molecules/cm³/s, dotted contours) calculated from the TOMCAT/SLIMCAT 3D chemical transport model.**

The largest ozone production rates are in the tropics where there is maximum solar insolation but the largest values for ozone concentration are at middle and high latitudes (Figure 21). The reason for this lies in the underlying stratospheric circulation. The Brewer–Dobson circulation is named after two scientists that helped to answer this ozone conundrum.

Rising tropical tropospheric air has low ozone. The ozone produced in the tropical stratospheric rises up the ascending branch of the circulation and towards the mid to high latitudes following an overturning cell structure. The ozone maximum occurs towards high latitudes in late winter and early spring, a direct result of the descending branch of the circulation. In contrast, there is little seasonal variation in the tropics. The method by which this circulation was discovered shows remarkable scientific ingenuity and insight (see Box 3).

Box 3 The global circulation of ozone inferred from aircraft observations over the UK

The story behind the reconciliation between theory and measurements begins in World War II. The UK military wanted to understand why aircraft that flew at high altitudes left condensation trails, an unwanted phenomenon that revealed aircraft positions to the enemy. Alan Brewer was assigned this task while he was working at the UK Met Office, supervised by Oxford Professor Gordon Dobson. Brewer needed accurate measurements of temperature and frost point at high altitudes, something only possible using the latest military aircraft.

Measurements from Brewer and other scientists showed that aircraft initiate the formation of a condensation trail by increasing the humidity within their exhaust trail. Long-lived condensation trails also require that the ambient air is supersaturated with respect to ice. These are conditions that we expect in the upper troposphere.

Brewer's major breakthrough came when he finally reached the stratosphere and found something remarkable. The temperature started to increase, and at these high altitudes no water was deposited and no condensation trails were formed. It was then clear that the reason why no condensation trails formed in the stratosphere was that the air was exceedingly dry. Indeed, the air was dryer than could be explained by atmospheric temperatures above the UK.

In 1949 Brewer then came up with an ingenious interpretation of this observed dryness at mid-latitudes. Dryness is maintained by a slow circulation of the air in which air rises at the equator through the tropical tropopause that is sufficiently cold to explain the dry air, moves poleward in the stratosphere, and then descends into the troposphere in temperate and polar regions.

(Cont.)

In the upper stratosphere, where the lifetime of O_x is short, ozone values predicted by the Chapman mechanism are much higher than observations: either the source of O_x is too large or the loss is too small. Chapman correctly predicted that ozone was produced by sunlight so there must be additional losses. This initially puzzled scientists because no other molecule species was sufficiently abundant to explain this additional loss. This flaw was not apparent until the 1950s when observations and knowledge of the chemical reaction rate constants reduced uncertainties such that they could no longer accommodate the discrepancy between model and data. It was only then that it was realized that catalytic cycles played a role in ozone in the stratosphere. The idea is summarized by a simple mechanism. Ozone reacts with species Z to produce ZO, which then reacts with O to reform Z. The net result is the loss of ozone to molecular oxygen, i.e. the same net effect as the Chapman scheme loss step:

$$(1)\ O_3 + Z \rightarrow ZO + O_2$$

$$(2)\ O + ZO \rightarrow Z + O_2$$

$$Net: O_3 + O \rightarrow 2O_2$$

The catalytic species Z is not lost but recycled to destroy other ozone molecules. Each cycle is initiated, has a propagation chain, and a termination point. During the 1950s it was discovered that ozone destruction could be initiated by water vapour, so that Z in that case is OH (hydroxyl radical). While it was a significant discovery it was not sufficient to reconcile theory and observations. The next breakthrough was due to Concorde. In the 1960s, supersonic aircraft were the latest technology.

With a London to New York transit time of less than three hours Concorde offered a very attractive prospect. To achieve these speeds Concorde cruised at an altitude of 18 km in the stratosphere where the air is thinner and consequently there is less air resistance. That also meant their exhaust was deposited in the stratosphere. This included nitric oxide (NO) from the oxidation of nitrogen at high temperatures. In the early 1970s, Paul Crutzen, while working at the University of Oxford reported that the catalytic cycle involving NO (Z in the above mechanism) was an effective sink of ozone. Most importantly, this cycle helped to reconcile Chapman theory with observations. In the end, for a number of reasons, relatively few Concorde aircraft flew (about thirteen out of a planned fleet of 500).

Shortly afterwards, Sherry Rowland and his researcher Mario Molina working at the University of California were investigating another catalytic cycle that involves chlorine from chloroflurocarbons (CFCs), which were used routinely as, for example, refrigerants and aerosol propellants. They are inert (lifetimes typically much greater than a century) until they reach the stratosphere where they release toxic chlorine. Later studies showed the importance of other halogens such as bromine.

In their forward-looking scientific paper, Molina and Rowland warned that increased use of CFCs (usage had increased 2–4 per cent since the 1930s) would seriously threaten the health of the stratospheric ozone layer. In 1985 a team of scientists at the British Antarctic Survey reported that springtime stratospheric ozone columns over their station at Halley Bay had decreased rapidly since the 1970s. This result was soon independently corroborated using some of the first space-borne observations of atmospheric composition.

Atmospheric models of stratospheric chemistry available in the 1970s and 1980s failed to predict the Antarctic ozone hole. At the time an important missing piece of scientific understanding was the role of polar stratospheric clouds (PSCs) in ozone destruction. These clouds, also known as nacreous or mother of pearl clouds, had been observed for over a century. They form at about 15–20 km above the Earth and they scatter light such that it gives them a pearly-white appearance. The stratosphere is dry so clouds dominated by water rarely form.

PSCs form at very low temperatures (less than $-80°C$) that are more often found in the Antarctic stratosphere than the Arctic counterpart, and can sometimes be seen over Edinburgh. PSCs come in a few types but the most prevalent involve nitric acid attached to three water molecules: nitric acid trihydrate, $HNO_3.3H_2O$. The surfaces of these PSCs help transform less reactive forms of chlorine into free radicals that subsequently catalyse the destruction of stratospheric ozone. Also, the formation of these clouds removes gaseous nitric acid that affects the chemistry cycles such that it increases ozone destruction.

The implications of the work done in the 1970s and 1980s demanded that the science and politics communities work more closely than they had previously. Progressively larger amounts of experimental evidence and the discovery of the Antarctic ozone hole led to a series of intergovernmental agreements (most notably the Montreal Protocol that was signed in 1987), which eventually led to the total ban on the production of CFCs and some other gases from 1996.

Hydrochlorofluorocarbons (HCFCs) were introduced in the 1990s as temporary substitute compounds for CFCs and added to the controlled substances in the Montreal Protocol. These compounds also contain halogens but their atmospheric lifetime is typically less than ten years so they do not reside in the atmosphere for as long as CFCs and can do less damage to stratospheric ozone.

We now see that increasing levels of HCFCs partially offset the decline in halogens from declining CFC levels. HCFCs are being gradually phased out over the next two decades.

Crutzen, Molina, and Roland received the 1995 Nobel Prize in Chemistry 'for their work in atmospheric chemistry, particularly concerning the formation and decomposition of ozone'. Research continues to fill the remaining missing gaps in our knowledge of the responsible chemical and physical processes associated with ozone.

Thankfully, we are now beginning to see the recovery of the reduced ozone concentration observed over the Antarctic during spring months. The recovery is consistent with theory but unequivocal detection of this recovery will take a few more years of measurements. A full recovery is expected by 2040, determined by the atmospheric lifetime of some of the gases involved. As the ozone-depleting gases emitted by humans decline we are now interested in the magnitude of natural gases that are emitted at the surface.

Troposphere

The troposphere, the lowest layer of the atmosphere, is where billions of people live and breathe. It is also where air pollutants are emitted, wildfires burn, vegetation grows, and where the oceans exchange gases. Tropospheric composition is an excellent example of where we did not initially appreciate the manifold consequences of human activities on the environment.

Historical context

Humans have influenced Earth's atmospheric composition for thousands of years from when we started to organize agricultural practices, for example flooding and draining of rice paddies, and raising livestock for consumption of meat and dairy. Whether these early activities, associated with far fewer people on Earth than today, imposed a significant imprint on climate records is still being debated. It is likely that natural processes mainly determined tropospheric composition before the Industrial Revolution that started in the late 18th century, which serves as the more established epoch for human influence on Earth's climate. This epoch marks the start of the anthropocene, a point in time when human activity began to be the dominant force shaping climate and the environment.

The beneficial impact of the Industrial Revolution on the world cannot be overstated, but an undesirable price for mechanized industrial processes overtaking hand production methods was unregulated emissions of pollutants and greenhouse gases. Large-scale coal mining was developed during this period to sustain the rapid economic growth, and coal provided the main domestic source of heat throughout the UK right up until the 1950s.

The environmental damage of coal first manifested itself as smog (a portmanteau of 'smoke' and 'fog'), especially over London. Given the numerous cultural references of London 'fog' by, for example, Dickens, Conan Doyle, Churchill, and Eliot, one can only conclude that residents simply resigned themselves to occasionally living in a thick, yellow-black smog. Smog was black and yellow because of soot particles from the incomplete combustion of the low-grade, sulphur-rich coal commonly burned by thousands of homes across the city, which mixed with the fog from the Thames Valley. The soot particles were accompanied by lethal concentrations of the poisonous gas sulphur dioxide. In December 1952, unusual anti-cyclonic conditions over London led to a temperature inversion, which prevented surface air being ventilated to higher altitudes. This resulted in a slow build-up of poisonous gases and soot particles that lasted more than a week. Over 12,000 people lost their lives due to that air pollution episode. One positive outcome of this event was the establishment of the 1956 Clean Air Act that banned the use of coal for domestic fires in urban areas. One result of this Act was the increase in the domestic use of cleaner sources of energy such as gas and electricity.

Unfortunately, we have not seen the last of coal-driven smog. Rapidly growing economies such as India and China often experience present-day high levels of pollution. Typically the underlying reasons are more complicated than for the London smog but the result is the same. These days we can observe the evolution of these events using space-borne sensors.

Meanwhile in 1950s Los Angeles, scientists were beginning to understand the origins of a different kind of smog that was brown and had an acrid smell unlike one produced by sulphur dioxide. Here, smog was linked with reduced visibility (similar to London), crop damage, eye irritation, and a strong smell. Car tyres wore out much more rapidly in Los Angeles than other cities due to the smog deteriorating the rubber. Arie Jan Haagen-Smit, a Dutch professor based in California, was the first to link the smog to vehicular emissions. Emission of nitrogen oxides (NO and NO_2) and hydrocarbons (literally chains containing hydrogen and carbon) react with sunlight to produce ozone and atmospheric aerosols (airborne particles); so-called photochemical smog. As we discuss in the section 'Present-day understanding', elevated concentrations of ozone and particulate matter (the particulate portion of an atmospheric aerosol) are detrimental to human health. It was the ozone that was cracking the rubber in tyres.

What followed (and continues today) is a concerted effort to regulate emissions of the ozone precursor gases. As it turns out it is also about 'location, location, location': the mountainous backdrop of Los Angeles that makes it such a beautiful city (on clear days) is what exacerbates the air pollution events by preventing ventilation of the sunlit cocktail of pollutants; similarly, Mexico City is surrounded by mountains and volcanoes and suffers terribly from poor air quality. Unlike the stratosphere, elevated concentrations of ozone in the troposphere should be avoided at all costs: good ozone versus bad ozone.

So much doom and gloom, but it is not all bad. The atmosphere has an ability to cleanse itself of pollutants. Up until the 1970s, the science community thought that on a global scale most tropospheric ozone was transported from the stratosphere. This was mainly because they thought that incoming photons were not energetic enough (after passing through the stratospheric ozone layer) to produce significant excited oxygen atoms to create hydroxyl radicals (OH). The global abundance of OH is typically equated to the oxidizing capacity of the troposphere. Scientists knew that pollutants such as methane and carbon monoxide (CO) reacted with OH so they argued that they must have long atmospheric lifetimes in the troposphere, with most of their destruction happening in the stratosphere. The only conclusion was that the world would soon face a catastrophic air quality problem due to the accumulation of pollutants.

In 1969, Bernard Weinstock from the Ford Motor Company used the ^{14}CO isotope, generated by cosmic radiation, to determine the lifetime of CO to be approximately one month. With this important work Weinstock helped 'to dispel further the concern that CO is accumulating in the atmosphere and represents a longtime hazard to human health' but was unable to confirm the removal process. Soon afterwards in 1971 Hiram Levy II from the Smithsonian Astrophysical Observatory showed that energetic photons do make their way down to the troposphere. This paper heralded the start of research into tropospheric chemistry as we know it today.

Nowadays, we have a reasonable idea of the atmospheric concentration of OH on a seasonal and hemispheric scale but detailed measurements are difficult to make because of its small concentrations and short lifetime (typically a second). Variations in OH are quantified using a halocarbon called methylchloroform (MCF) that was phased out by the Montreal Protocol. Emissions of MCF, associated with its industrial application as a solvent and degreasing agent, are reasonably well known. Because its only tropospheric sink is OH by looking at MCF variations in the troposphere we can infer the OH distribution.

Present-day understanding

As we understand it today tropospheric ozone is determined by a balance of two large, similar values for its production and loss, and a smaller source from the transport from the stratosphere. The source from *in situ* production is of the order of 5000 Tg O_3/year (a teragram is one million tonnes) and the source from stratospheric transport is 500 Tg O_3/year. The *in situ* loss is of the order of 4500 Tg O_3/year and deposition to the land and ocean surface is of the order of 1000 Tg O_3/year. Even though this stratospheric source is relatively small it does affect ozone variations throughout the troposphere. Figure 22 shows a simple schematic that outlines the basics of ozone chemistry in the troposphere.

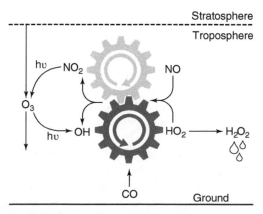

22. Schematic of tropospheric ozone chemistry.

Ozone in the troposphere is determined by transport of ozone from the stratosphere and the *in situ* production and loss in the troposphere. Within the troposphere, ozone is split by photolysis into molecular oxygen and an excited oxygen atom. The excited oxygen atom reacts rapidly with water vapour to produce two OH radicals. OH is the main tropospheric oxidant that is responsible for the removal of many pollutants and greenhouse gases.

Removal of OH from the troposphere is by reaction with carbon monoxide (CO) and hydrocarbons. To illustrate the chemistry I will focus on CO. Oxidation of CO by OH produces an excited hydrogen atom that reacts with molecular oxygen to produce the hydroperoxyl radical HO_2.

The fate of HO_2 depends on the abundance of nitric oxide (NO) in the atmosphere. If NO is relatively low compared to HO_2, then HO_2 radicals can react with each to produce hydrogen peroxide H_2O_2. Hydrogen peroxide is highly soluble and has an atmospheric lifetime of the order of a week due to precipitation. If NO is relatively high then HO_2 can also react with NO to produce OH and nitrogen dioxide (NO_2).

Nitrogen dioxide can be photolysed to return NO and an excited oxygen atom. Now we have come full circle. The chain mechanisms are the two cogs that cycle between OH and HO_2 and between NO and NO_2, which are initiated by the source of OH from the reaction of water vapour and the excited oxygen atom. In the troposphere, HO_x (sum of OH and HO_2) and NO_x (sum of NO and NO_2) catalyse production of O_3. In the remote troposphere, where NO is very low, O_3 can be lost by reaction with HO_x.

Tropospheric NO_x is mostly due to fossil fuel combustion and biomass burning (e.g. deforestation and agriculture), but also from soil (from microbial processes), lightning (very high temperatures can split molecular oxygen that can be react with nitrogen N_2), and the oxidation of ammonia (NH_3). Reactive nitrogen (NO_y, which is the sum of NO_x and its oxidation products) can be temporarily locked up in reservoir species that are more stable. An example of this is the formation of peroxyacetyl nitrate (PAN), which is more stable at cooler temperatures, so once it is lofted out of the boundary layer it has a lifetime of many weeks. But once air laden with PAN begins to descend into the warmer, lower troposphere it releases the nitrogen species. This means that the effective footprint of a continental pollutant source can reach thousands of kilometres downwind and sometimes affect otherwise clean marine environments (see Box 4).

Box 4 Downwind Consequence of Air Pollution

Acid rain is an environmental consequence of uncontrolled emissions of air pollutants that are transported far downwind from the source region. This only came to prominence in the 20th century but was first documented in the 19th century. In 1872 the Scottish chemist Robert Angus Smith published a small book called 'Air and rain: the beginnings of chemical climatology' that reported high levels of acidity in rain falling over industrial regions in England and lower values over less-polluted coastal regions. Little attention was paid to this book until the 1950s when biologists noticed an alarming decline in fish populations in southern Norwegian lakes with similar findings over North American lakes in the 1960s. They found that the rain contained high concentrations of nitric and sulphuric acid due to the oxidation of nitrogen oxides and sulphur dioxide (SO_2) emitted by fossil fuel combustion. The puzzle: SO_2 typically could survive in the atmosphere without being oxidized to sulphate for a couple of weeks, which is plenty of time for it to be transported far downwind. But sulphate in rain was highest where SO_2 was being emitted. The answer to the puzzle did not appear until the 1980s when it was found that this chemical transformation was acid catalysed. In America this led to several amendments of the US clean air act to reduce progressively emissions of SO_2 from power plants.

The most abundant hydrocarbon is methane (CH_4) whose sources are evenly split between human activity (anthropogenic) and nature (biogenic). Higher hydrocarbons (containing greater numbers of hydrogen and carbon atoms) are typically dominated by either anthropogenic or biogenic sources. The most abundant biogenic hydrocarbon is isoprene (C_5H_8) that is emitted by certain (broadleaf) trees, shrubs, and grasses during hot and sunny weather. The reasons why plants emit isoprene are not obvious. Why would they voluntarily lose useful carbon? One widely accepted hypothesis is that vegetation releases isoprene as part of a mechanism to protect itself against hot temperatures.

Questions still remain about the role of isoprene emission as a mechanism to improve tolerance to ozone and other reactive oxygen species that at elevated concentrations can damage plants. In any case, O_3 chemistry does not care about the source of hydrocarbon fuel and when it was pointed out that on a regional basis biogenic hydrocarbons could actually compete with (and in some instance dominate over) anthropogenic hydrocarbons it was met with glee in some political circles, particularly in the United States during the Reagan era where it was suggested that we should hold off introducing tough emission standards for anthropogenic sources.

There are many hundreds of biogenic gases that are emitted by flora stimulated by plant growth, development, reproduction, and defence. Plants can defend themselves against predators and can communicate with insects (pollinators) and other plants. We are largely unaware of it but the atmosphere is awash with chemicals that correspond to what are essentially plants nattering amongst themselves. Much remains to be learnt about this rich and fragrant part of Earth's atmosphere.

The ocean is also biologically active, which results in a wide range of gases being emitted. Hydrocarbons, oxygenated compounds, and halogens are all emitted in quantities that vary with season and location. The magnitude of the net emission to the atmosphere relies on the concentration gradient between the ocean and the atmosphere and also wind speed that provides the mechanism to speed up the exchange. Data from flights over the remote Pacific, for example, show that emitted gases are sufficient to maintain a reactive chemical environment.

As we reduce emissions of halogens from anthropogenic sources we are progressively vulnerable to the remaining natural sources. Despite some of these having atmospheric lifetimes of days to weeks, we find they make their way to the upper tropospheric lower stratosphere due to rapid, large-scale convective cells that are found over specific geographic regions, e.g. the western Pacific (Chapter 3).

Because of the interrelations between the chemicals associated with ozone chemistry there is a fine line between ozone production and loss. Using models that incorporate the thousands of chemical reactions we can begin to understand how a decrease/increase in hydrocarbons/nitrogen oxide emissions can change ozone. Depending on where you are on this map a well-meaning decrease in hydrocarbon emissions might not change ozone. This illustrates why a solution in one city might not necessarily translate to a similar reduction in pollution in another. It also illustrates that only a systemic view on air quality will lead to successful outcomes. For example, the UK government in 2010 encouraged UK car owners to buy diesel vehicles because they emitted less carbon dioxide. However, they emit more nitrogen oxides and particulate matter, with the associated implications for human health.

Atmospheric aerosols

Aerosols are small particles or liquid droplets suspended in air or another gas. Particulate matter refers to the particulate portion of an aerosol. Primary (direct) sources of aerosols include the mechanical action of wind eroding surfaces and lifting small particles, incomplete combustion, and biological aerosols (e.g. fungal spores, pollen, viruses). Secondary aerosols comprise low-volatility gases that condense on existing aerosols. Aerosols range in size from a billionth of a metre (nanometre) for those generated by gas condensation to a few microns for those generated by aeolian processes such as mineral dust and sea salt. They are important because as we have already discussed they scatter and absorb incoming solar radiation that can affect visibility and the radiative properties of clouds; elevated concentrations are associated with respiratory illnesses; and they provide small surfaces for chemical reactions that would not take place otherwise. Figure 23 illustrates the aerosol life cycle.

Emissions that include precursor gases will eventually condense onto existing particles or primary particles. That nucleation process leads to the formation of ultrafine aerosols that are small, typically submicron. The next process is condensation or coagulation of aerosols that results in aerosol growth. Depending on the chemical composition these aerosols might be scavenged or evaporate via cloud–aerosol interactions. Larger aerosols (super-micron) are more likely to be lost by deposition. Coarse aerosols can be mechanically emitted directly into the atmosphere.

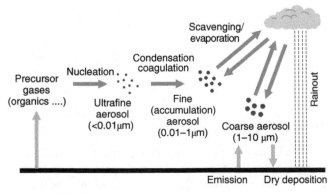

23. **Aerosol life cycle.**

The residence time of an aerosol depends on the size (as a proxy for mass) of the aerosol. For very ultrafine aerosols this value is typically less than a day before they coagulate to form larger particles. Similarly, for large aerosols the residence time is about a day due to removal by sedimentation and precipitation scavenging. Accumulation aerosols that have a diameter between a fifth and two microns have a residence time of up to 1 to 2 weeks in the upper troposphere and a few days in the mid and lower troposphere.

Within a few days, aerosols can be transported thousands of kilometres before making contact with the ground again. Black carbon aerosol, for example, emitted from incomplete combustion can be transported far from its source to pristine, snow-laden surfaces and decrease the surface albedo resulting in surface warming (Chapter 2).

Aerosols grow according to their ability to absorb water. Water-soluble aerosols start to increase and decrease in size at a certain value of relative humidity. The deliquescent point determines the transition from solid to liquid and the efflorescence point determines the transition from liquid to solid. These phase transitions depend primarily on the chemical composition and thermodynamics.

Aerosol size matters for a number of reasons. Smaller particles tend to settle further into your body and cause more serious health problems. Larger particles can typically be coughed up—inspecting the contents of a tissue after blowing your nose while in any large city will provide evidence that your body filters particulate matter. Light will scatter preferentially at wavelengths that are close to the size of the aerosols.

Aerosols are composed of a mix of organic and inorganic material. Inorganics include sulphate, nitrate, ammonium, and chloride. A key precursor gas is sulphuric acid, from the oxidation of sulphur dioxide from fossil fuel combustion and other sources, which can condense under atmospheric conditions to form aqueous sulphate particles. The sources for the organic material are too numerous to list but the mixture you ingest will depend on whether you are in, or downwind of, an urban environment or in a remote environment. For instance, urban Edinburgh aerosols have a typical mass concentration of 3 $\mu g/cm^3$, comprising approximately two-thirds from organic material. Aerosols from a rural forest in Finland have a typical mass concentration of 2 $\mu g/cm^3$, and similar fractional contributions of organic and inorganic compounds to Edinburgh. The main difference is in the organic fraction. In Finland, the organic fraction has a large contribution from oxygenated organic aerosol from biogenic sources.

Greenhouse gases

The two remaining elephants in the room are carbon dioxide and methane. There are many more elephants in the herd but here we focus on the two largest, noisiest ones. Figure 24 shows what is now considered an icon of the 20th century: the Keeling curve named after Charles Keeling who initiated measurements at Mauna Loa in 1957 as part of the International Geophysical Year.

Carbon dioxide is now accumulating in the global atmosphere at about 2 ppm/year, which equates to 4 billion tonnes of carbon per year. Atop of the linear increase in atmospheric carbon dioxide is also a seasonal cycle that reflects the uptake of carbon dioxide by terrestrial vegetation during the growing season and the net release of carbon dioxide from the biosphere due to microbial decay outside the growing season; this seasonal cycle is reversed in the southern hemisphere.

Even within a few years of these data being collected it was clear that fossil fuel combustion and land use change were imprinting global signatures on carbon dioxide concentrations in the remote atmosphere, something that had previously not been fully

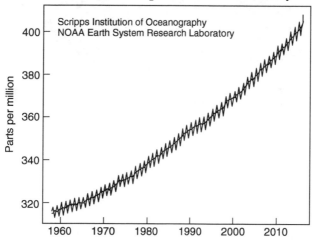

Atmospheric CO₂ at Mauna Loa Observatory

Scripps Institution of Oceanography
NOAA Earth System Research Laboratory

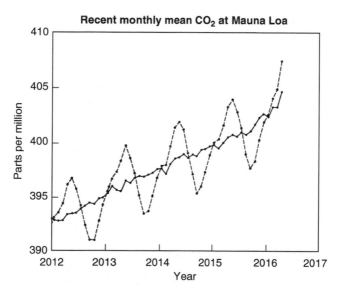

Recent monthly mean CO₂ at Mauna Loa

24. Atmospheric variations of carbon dioxide and CH_4.

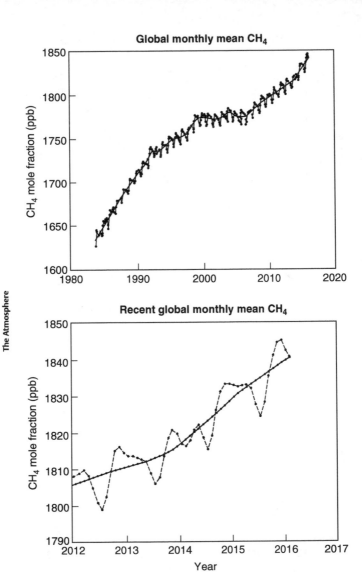

24. Continued

appreciated although some pioneers had hypothesized this idea many years before. These data also showed the natural breathing of the planetary biosphere. The depth of the seasonal cycle and the distance between successive cycles provides us with clues about the biological activity in a particular year and the length of the growing season, respectively.

There are still many gaps in our understanding of changes in atmospheric carbon dioxide over decadal to centennial timescales but one is larger than most with far-reaching scientific and policy implications. Figure 25 crudely summarizes our current understanding of the global carbon cycle. We have excellent knowledge of how much carbon dioxide is in the atmosphere and its accumulation rate. On a global scale, the largest source of carbon dioxide in the atmosphere is from fossil fuel combustion and cement production, which we regulate heavily so we think we know how much is emitted. Land-use change is a minor and uncertain source.

The two main losses of carbon dioxide from the atmosphere are uptake by the land and ocean biospheres, with geological

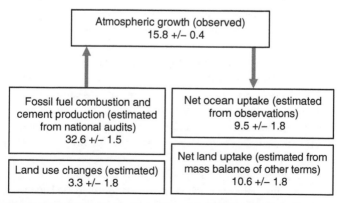

25. Present-day understanding of the global fluxes of carbon dioxide (Pg C/year).

losses acting on much longer timescales. No single atmospheric lifetime for carbon dioxide can be given because of the different processes involved that act on vastly different timescales. A simple back-of-the-envelope calculation is all that is needed to reveal that there is discrepancy between what we know is being emitted, what we know is staying the atmosphere (the growth rate), and our best knowledge of the surface processes that absorb atmospheric carbon dioxide. This is sometimes called the 'missing sink' problem—but we are not missing it, we just have not yet confirmed its whereabouts.

Even this level of uncertainty has only been possible since high-quality observations of carbon dioxide started in the late 1960s. It seems odd to us now that it took so long to observe this important gas, but hindsight is 20:20. What we do know is that a remarkably constant 44 per cent of annually emitted carbon dioxide remains in the atmosphere. Because the absolute amount of carbon dioxide emitted into the atmosphere increases per year this means that the uptake processes are increasing to match the amount of carbon dioxide in the atmosphere.

Methane is both chemically reactive and radiatively active. On a molar basis it is some twenty-five times more potent than carbon dioxide at trapping heat in the atmosphere. It has an atmospheric lifetime of approximately nine years so while emissions are increasing it does not accumulate in the atmosphere like carbon dioxide. The largest sources of methane are from wetlands (from microbial processes), fossil fuels, ruminants such as cows and sheep, fires, landfill, and rice cultivation, with minor sources from termites and oceans. The main loss process is from the oxidation by OH with a smaller source from surface soil uptake.

Figure 24 shows that over the past 50–60 years, methane concentrations in the atmosphere have generally increased on an annual basis. The only major exception is a 7-year hiatus between 2000 and 2006 when the annual increase in atmospheric methane was essentially zero. This is an example of mass balance when production and loss of methane remain large but they balanced each other on a global scale. One of the likely culprits for some of these changes is the emissions from wetlands that are responding to unusual temperatures at high latitudes and anomalous precipitation at low latitudes, although this is still debated by scientists. Atmospheric methane plays a role in climate and atmospheric chemistry, and studies have shown that targeted reductions in methane emissions would reduce tropospheric ozone and have a positive effect on climate via radiative forcing.

Upper atmosphere

Chemical processes also occur in the mesosphere and thermosphere. This is one of the harder-to-reach parts of the atmosphere, with measurements limited to instruments launched by rockets, ground-based radar, and a few satellite sensors, but it is a part that contains some of the most breathtaking natural wonders.

In the mesosphere, where the atmosphere mainly consists of oxygen atoms and nitrogen molecules, ultraviolet solar radiation breaks apart molecular oxygen resulting in excited oxygen atoms. When these excited atoms eventually return to their normal state they emit a tiny amount of infrared radiation that is visible to the naked eye. This phenomenon is called airglow and manifests itself as a faint green light from the upper atmosphere, which is due to oxygen atoms between 90 and 100 km up in the atmosphere. There is also a weaker red light from oxygen atoms higher up in the atmosphere, where these gases are less dense and collide less frequently. Other gases that contribute to the glow include sodium and nitrogen. The glow is comparable to the light from a candle 100 m in the distance, but it is routinely observed by a camera aboard the International Space Station.

Aurorae are a more dramatic version of the same excitation process, driven instead by charged particles from the Sun. In the northern hemisphere, it is called the Aurora Borealis (northern lights) and in the southern hemisphere it is called the Aurora Australis (southern lights). These charged particles, mainly protons and electrons travelling at supersonic speeds, vary in strength due to solar activity and carry a weak solar magnetic field called the interplanetary magnetic field. Earth is shielded from a direct onslaught of this solar wind by a magnetic field that is generated by a dynamo from the rotating, turbulent molten iron in its outer core.

The resulting interaction of the solar wind and magnetic field causes a teardrop distribution of magnetic lines, which prevents substantial leakage of these particles in Earth's atmosphere. However, during periods of elevated solar activity the solar wind can interact strongly with Earth's magnetic field. The subsequent reorganization of the magnetic lines can link the fields of the Sun and Earth, allowing solar particles to flow inside Earth's atmosphere forming a ring around each magnetic pole. It is these high-speed particles that interact with the gases in the upper atmosphere. The most common Auroral colour is a green because of the oxygen content of the upper atmosphere, as described earlier. Red auroras are due to oxygen atoms over 300 km in the atmosphere.

Chapter 5
Atmospheric measurements

Measurements of the atmosphere improve the performance of
weather forecasts, underpin our understanding of Earth's
changing environment, and help inform climate models that are
used to look at future climate scenarios.

Why measure?

Measuring the atmosphere is non-trivial and expensive so why
do we bother? We have mathematical models that describe
atmospheric phenomena that have some predictive capability so
surely these will suffice?

There are two flaws to this line of reasoning. By ignoring
measurements we are implicitly assuming that all processes are
understood and that our climate models can reproduce the world
perfectly. This is of course nonsense. There is plenty of science still
to be discovered and/or refined about the atmosphere and how it
interacts with Earth's surface and the Sun. Humans also have a
role to play as they alter surface landscapes and consequently
affect directly and indirectly Earth's atmospheric radiation budget.
To quote President John F. Kennedy's famous 1962 speech at Rice
University: 'The greater our knowledge increases, the greater our
ignorance unfolds,' a sentiment that is sometimes forgotten.
We are destined forever to explore deeper into how Earth's

atmosphere works without necessarily reaching knowledge that is sufficient to address ever more pressing (and sometimes emerging) challenges.

The other main difficulty lies in the non-linear relationships between variables: a unit change in one independent variable might result in a disproportionately higher or lower change in other dependent variables. Chaos theory tells us that the smallest error in knowledge of the independent variable can disproportionately affect the dependent variables. In the worse case scenario the initial error can grow rapidly as it is propagated in time and space such that the resulting model describes a completely different atmospheric state that might bear little or no resemblance to the true atmospheric state.

The classic metaphorical example of this is the so-called butterfly effect; a name coined by Edward Lorentz, an MIT professor and pioneer of chaos theory, in the 1960s. The formation and path taken by a hurricane is influenced by minute perturbations to the atmosphere such as those from the flapping wings of a distant butterfly weeks earlier.

So we need measurements to 'correct' models so they do not stray too far from the truth. This correction can take the form of nudging the atmospheric state towards the observed state or using data to improve the processes within the model that describe the atmospheric state. Measurements are also imperfect, so we cannot simply assume they are the truth: the outcome is a merging of the information provided by the models and measurements with the understanding that combining information will produce a more realistic result than using either set of information.

Horses for courses

Generally many different types of measurements are required to sample the atmosphere. Each type has its own strengths and

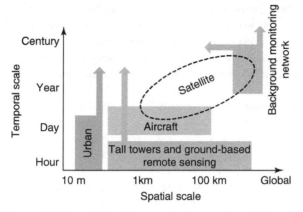

26. Characteristic spatial and temporal scales associated with different ways to measure the atmosphere.

weaknesses (e.g. Figure 26) and only by weaving together information from these different types can we afford a more complete picture of the atmosphere than we would otherwise have from any individual source. Measurements collected in the urban environment, e.g. roadside measurements of temperature and air pollution, tend to represent variations over a small spatial scale but can be collected over long periods to perhaps discover something about seasonal and annual changes.

To understand changes in the atmosphere from many kilometres away you have to sample air far above the surface. A tall tower that is a hundred metres above the ground will sample air that is indicative of a spatial area that is perhaps hundreds of square kilometres. In many countries, scientists use telecommunication masts that can be 100–200 metres in height, and collect air near the mast tops by pumping air down through Teflon tubes to an analyser housed near the base of the mast. These types of measurements are also typically run for many years. A similar measurement can be part of a monitoring network. The main difference is that the locations of these monitoring sites are far from local sources (e.g. cities, forest) so that the measurements are representative of the background atmosphere, i.e. very large (hemispheric) spatial scales. The value in these measurements is also by virtue of their longevity, so that they can be used to understand how climate affects atmospheric composition.

Aircraft measurements are typically collected during a campaign of a day to a few weeks, but they provide invaluable vertical information and can sample spatial scales that range from one hundred metres, say, to hundreds of kilometres. Lastly, space-borne observations provide a global view on the atmosphere and can begin to link measurements collected at smaller spatial and temporal scales. The typical lifetime of a satellite mission is typically a few years but they can last for much longer.

Harnessing the information described by these different observations requires a model and some statistical machinery that can harmonize the model and data, taking into account their respective errors. This is a big data challenge that involves large and heterogeneous streams of information that need to be stitched together, often by major weather forecast centres, to generate the most impactful scientific results.

Measurements from below

The first atmospheric measurements were taken from the land surface hundreds of years ago. The 15th century saw the introduction of measurements of rain, wind, and humidity, with temperature and pressure measurements introduced in the 17th century. The motivation for these measurements was not always scientific curiosity.

A standardized rain gauge was introduced by the reigning dynasty in 15th-century Korea in order to quantify rainfall that could then be equated with potential agricultural harvest

so that farmers could be taxed accordingly. This was some
200 years before the polymath Sir Christopher Wren invented the
tipping-bucket rain gauge.

The German physicist Daniel Fahrenheit introduced the mercury
thermometer in the early 18th century, and developed the
temperature scale still used in some parts of the world. A few
years later, the Swedish astronomer Anders Celsius introduced an
alternative centigrade-based temperature scale. But the absolute
measure of temperature is the Kelvin scale, which is maintained
by the International Committee for Weights and Measures.
The kelvin is defined at the 1/273.16 of the thermodynamic
temperature that defines the triple point of water, where gas,
liquid, and solid coexist.

The technology to measure atmospheric temperature has not
changed much over the past hundred years. One innovation was
to cover the thermometer from direct sunlight and also from
inclement weather. You may well have seen these Stevenson
screens: a white box typically 1.25 m above the ground that takes
advantage of natural ventilation using horizontal slats. The
screens were named after Thomas Stevenson, a 19th-century
British engineer who was also father to Robert Louis Stevenson.
Within modern-day Stevenson screens, Mercury thermometers
have now been replaced by thermometers that determine
temperature by measuring the electrical resistance of a pure
platinum wire, which has the advantage of being unreactive and
the results being reproducible. The tipping-bucket rain gauge
is still used widely to measure precipitation.

During my (pre-internet) childhood I had a mercury barometer
hanging on the wall at home. A small glass tube filled with
mercury would rise and fall with atmospheric pressure. If the
pressure was high we could look forward to fair weather and low
pressure was associated with unsettled weather (a euphemism for

cloud, rain, and generally miserable weather). Of course, accurate measurements of atmospheric pressure are of fundamental importance in weather prediction. Modern-day barometers that feed these predictions use sophisticated sensors that generate repeatable measurements and are stable over long time periods. A typical instrument contains more than one sensor to ensure reported values are not erroneous.

Nowadays, weather stations that measure a large and wide range of variables have replaced sites that collect only a few measurements. A typical site might include atmospheric pressure, temperature, humidity, wind, precipitation (for example, showers with/without thunderstorms, frozen precipitation, rain, drizzle), visibility, sunshine, radiation, and cloud properties (type, coverage, base height, and temperature). These kinds of sites are placed across the UK and elsewhere, spaced apart by tens of kilometres or so with the intention that they capture variations over a (synoptic) spatial scale of a typical frontal system. These are being used on a global scale to observe changes in surface temperatures for weather forecasting and for studying links with changes in climate. A great deal of effort goes into locating these measurements so they are not biased by being on a hill or shaded by trees or being heated by nearby buildings or on the ground. This can mean the difference between, for example, data supporting the idea of a slowdown in global warming or not. Where and how you measure is vitally important.

Measurements of trace gases started in the late 19th century with ozone atop mountain sites. Data collected at Pic du Midi Observatory in the Pyrenees (3000 m above ground level) using a simple method developed by the German chemist Christian Friedrich Schönbein who had isolated ozone in the mid-19th century. He exposed paper that was soaked in a mixture containing potassium iodide and starch to atmospheric ozone. Ozone in the air oxidized the potassium iodide to produce iodine, which then reacted with the starch to stain the paper a shade of purple: a darker purple indicated more ozone present in the atmosphere.

Ozone was measured initially to support medical applications. In the 19th century medical thinking suggested that ozone was good for the lungs and provided a cure for the likes of tuberculosis, bronchitis, and laryngitis. Ozone was even used in World War I to disinfect wounds. Nowadays, ozone therapy is no longer considered effective. These early mountaintop measurements are crude by today's measurement standards but despite uncertainties the rise in ozone recorded in the Pic du Midi observations are similar to concurrent values measured in Croatia and Italy. They offer us a record of how humans have altered surface air pollution over the past century.

Similarly long records for other chemical species are few and far between so our historical knowledge of atmospheric composition, including ozone, on a global scale is often limited. The Keeling curve based on atmospheric carbon dioxide measurements at Mauna Loa in Hawaii, described in Chapter 4, is an exception. These measurements provided the world with the first evidence that humans had altered the atmosphere on a global scale—something that was not fully appreciated before the measurements had been collected. Nowadays, Mauna Loa is one of a hundred or so sites that constitute a network of measurements that are carefully calibrated on the same measurement scale so we can compare measurements collected at different stations.

A common measurement technique is gas chromatography to separate individual chemical compounds coupled with a detector. A common detector for atmospheric composition is a mass spectrometer because it is sensitive to small quantities, but other detector techniques are also widely used. Other measurement networks using a variety of instruments have been established to address different scientific goals, e.g. to support the understanding of the long-term evolution of the stratospheric ozone layer or to help validate space-borne observations of carbon dioxide. The one thing they have in common is that a tremendous amount of care is given to ensure inter-comparability of sites by establishing a reference standard for a gas that everyone can use to test that their equipment reproduces a particular value. It is the measurement inter-comparability that gives the network real scientific value.

Ground-based observations that are fixed in location, such as those collected within a network, provide a measurement of an air parcel that passes over the instrument (Figure 27). This is called the Eulerian frame of reference. A measurement taken in this framework provides a snapshot of an air parcel in time and space. The complementary Lagrangian frame of reference moves with the general flow of air so that it can be used to understand how an air parcel evolves thermodynamically and chemically in time and space. In practice, Lagrangian experiments are difficult to achieve because they require an accurate estimate for the transport of an air parcel.

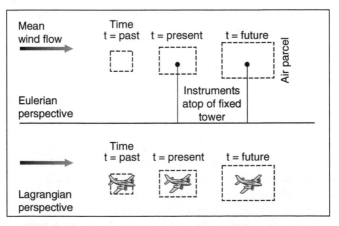

27. **Schematic describing the Eulerian and Lagrangian perspective of the atmosphere. The progressively diluted air parcel is denoted by the dashed rectangles.**

Measurements within the atmosphere

The weather balloon has been the workhorse for meteorological measurements, and continues to play an important role in weather prediction. These balloons (which, together with their instruments, are better known as sondes) are up to 1.5 metres in diameter when they are launched and slowly expand as atmospheric pressure falls off with altitude (Chapter 2). They are filled with helium and rise rapidly to the tropopause carrying small amounts of expendable instrumentation that can collect information about temperature, humidity, and winds. As they ascend the instruments relay measurements to a ground station. When the balloon reaches a certain altitude it bursts and the instruments are returned to Earth using a small parachute.

Atmospheric research campaigns use much larger balloons (with volumes ranging from 300,000 to over a million cubic metres) that are capable of carrying much heavier scientific instruments (a few metric tonnes) to the stratosphere. Launching instruments on rockets are less common but they can reach the mesosphere (Chapter 1). At one point in time, measurements collected by sondes produced the only data we had about the atmosphere above the surface. Nevertheless, using these sparse data scientists were still able to infer changes in general circulation.

Research aircraft are essentially airborne laboratories. These are aircraft that you might use for regional/national flights but with the chairs removed and replaced by racks of instruments. Metal tubes and inlets are fitted to the outside of the aircraft that suck air into pipes that pass inside the cabin and get siphoned off to a range of instruments to measure different attributes of gases and aerosol particles (Figure 28). There are also instruments mounted in wing pods that measure properties of cloud particles. Other *in situ* instruments provide meteorological data. The advantages of using an aircraft is that it can provide fine-scale

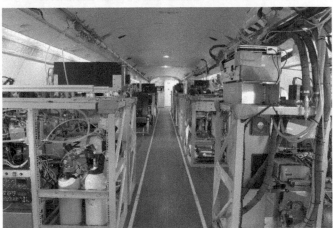

28. The UK BAe-146 atmospheric research aircraft: (top) taxiing on the runway with air inlets visibly protruding from the fuselage; and (bottom) the racks of instruments inside the main cabin.

information in the vertical dimension that is impossible to achieve using ground-based instruments. An aircraft can also move with the general flow of air so that it can be used to study the atmosphere using the Lagrangian frame of reference (Figure 27).

Aircraft allow scientists to study certain parts of the global atmosphere that are difficult to access on the ground, e.g. over remote terrestrial and marine ecosystems, and have the space to provide a more comprehensive description of atmospheric composition than can be achieved using a sonde. They also measure properties of cloud, such as the formation of ice, which are usually not possible using ground-based instruments. Most aircraft will stay below the stratosphere in the troposphere. In collaboration with some major commercial airliners, we now routinely have a small suite of atmospheric measurements limited to cruising altitudes (approximately 10 km above the surface) along major international air corridors and profiles during take-off and landing.

Aircraft observations of the stratosphere are difficult but they are possible. NASA previously used their U-2 aircraft and later their ER-2 aircraft, which can reach altitudes higher than 20 km, to collect data that eventually underpinned our understanding of stratospheric ozone. The U-2 aircraft is an ultra-high reconnaissance aircraft that was flown during the Cold War, which was retrofitted with atmospheric science instruments. NASA has recently added repurposed ex-military Global Hawk reconnaissance unmanned airborne vehicles (UAVs) to its high-altitude aircraft fleet. These UAVs are not limited by human endurance and can fly for over a day, allowing scientists to explore more remote parts of the atmosphere. We are only beginning to understand how best to use these data, particular in conjunction with other conventional platforms.

Smaller-scale UAVs such as fixed-wing (planes) and rotary (copters) also have a role to play in studying the atmosphere, and they are becoming more common in the scientific community. They have the advantage that they can be deployed quickly with small, light payloads to survey dangerous or undesirable environments closer to the ground such as over landfills, volcanoes, and hurricanes. As the UAV and measurement technology improves these platforms will be able to lift more mass for longer periods of time.

Measurements from above

The first artificial Earth-orbiting satellite was Sputnik 1 launched by the Soviet Union in the International Geophysical Year of 1957. The size of a beach ball and weighing more or less the average UK male weight of 84 kg, Sputnik 1 had far-reaching military, political, and scientific implications and heralded the start of the Space Race. The drag of the spacecraft as it orbited between 215 and 950 km above Earth's surface provided information about the density of the upper atmosphere, and the infamous intermittent bleeps received on Earth from the telemetry provided information about the propagation of radio waves through the ionosphere. More importantly, it heralded Earth observation from space. Explorer 1 was launched by the US in 1958 as part of the International Geophysical Year and detected the Van Allen radiation belts (Chapter 1).

We have made rapid progress since the heady days of the 1950s. Nowadays, there are nearly 200 Earth-observing satellites that measure meteorological variables (e.g. temperature, wind, clouds, humidity), radiation, chemical composition, land surface properties, and geophysical properties (e.g. gravity, magnetic field). These satellite data now form the basis of detecting changes in regional temperature and precipitation that can be linked to changes in climate.

There are two main methods that satellite instruments use to sample the atmosphere. Passive remote sensing instruments measure radiation that has passed through Earth's atmosphere where it is absorbed, reflected, or scattered by atmospheric composition, clouds, and the surface. Active remote sensing instruments provide their own energy to probe the atmosphere: a pulse of energy is fired down through the atmosphere and the returned energy that is collected by the instrument provides information about the atmosphere. There are advantages to both methods but the passive technology is more widespread. Both methods rely on the propagation of radiation through the atmosphere, which we have discussed in Chapter 2.

There are several geometries that satellites use to remotely sense the atmosphere on Earth and elsewhere in the solar system, depending on the scientific questions that are being addressed. The two most common approaches are to observe the atmosphere using limb sounding or nadir sounding (Figure 29). The two variants of either of these two approaches depend on whether the satellite is observing thermal infrared or microwave radiation emitted directly from the surface and atmosphere or observing shorter wavelengths in which case the Sun provides the source of energy that is observed as it passes through the atmosphere. Both approaches are complementary and often limb sounders and nadir sounders can be found on the same satellite.

Generally speaking, nadir sounders provide vertical resolution of the order of a kilometre for temperature and humidity (and column integrated amounts for trace gases) and have a horizontal ground dimension of the order of 10 km. Although limb sounders generate measurements that can have sub-kilometre vertical resolution in the stratosphere and lower troposphere, as the limb gets progressively closer to Earth's surface it also gets longer so that the horizontal resolution of each successive measurement becomes coarser, increasing to several 100s km.

There are also several orbits that a satellite can use to probe the atmosphere. The most popular orbits used by the atmospheric community are geostationary and Sun-synchronous orbits.

A satellite in a geostationary orbit sits at 35,800 kilometres above Earth's equator. At this altitude the orbital period of the satellite is the same as the Earth so it appears to hover over the same

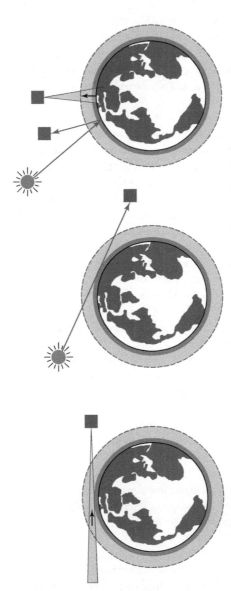

29. Schematic that illustrates some common measurement approaches used by satellites. The left and centre plots show limb sounding of the atmosphere using thermal/microwave and solar radiation, and the right plot shown nadir sounding of the atmosphere using thermal and solar radiation. The depth of the atmosphere is exaggerated.

geographical region. This type of orbit is often used by meteorological centres to observe weather patterns and is also used by telecommunications satellites. The wavelength ranges used by meteorological sensors to measure temperature, clouds, and humidity often allow scientists to observe thermal anomalies associated with fires and to observe dust storms that are a seasonal feature downwind of desert regions. We will soon have air pollution sensors sitting in this orbit, allowing scientists a first glimpse of the diurnal variation of atmospheric chemistry and surface air pollution concurrently on spatial scales ranging from 10 to 1000 kilometres.

A satellite in a Sun-synchronous orbit typically sits below 1000 km with a precession rate that exactly matches the time it takes for Earth to orbit the Sun. The result is a satellite that samples the atmosphere at the same local time of day throughout the year, twice a day separated by 12 hours, e.g. 09.30 and 21.30. Scientists often favour this orbit because it keeps the angle of the sunlight on Earth's surface as consistent as possible throughout the year that might otherwise confuse interpretation of observed atmospheric changes.

In the 1970s satellite technology was being developed to address the growing needs of the meteorological community to support weather forecasting, but also to address the growing awareness of large-scale changes to stratospheric ozone (Chapter 4). In late 1978 NASA launched the first of many Total Ozone Mapping Spectrometers (TOMS). The TOMS satellites played a vital role in verifying the persistence and spatial extent of the depletion of stratospheric ozone over the Antarctic. Data from TOMS and other space-borne instruments help also to understand the underlying reasons for the magnitude and variation for these changes in ozone.

The first inkling of the extent of atmospheric pollution in the troposphere was in the 1980s when NASA launched a simple

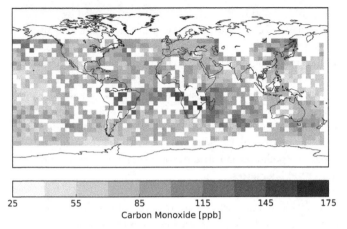

30. Distribution of carbon monoxide collected by the NASA MAPS instrument aboard Space Shuttle Columbia, October 1984.

Carbon Monoxide [ppb]

25   55   85   115   145   175

Earth-observing instrument, MAPS (Measurement of Air Pollution from Satellite), that could measure carbon monoxide as a science payload on the Space Shuttle Columbia, on the ill-fated Space Shuttle Challenger, and twice in the mid-1990s on the Space Shuttle Endeavour. The distribution of carbon monoxide from the instrument was patchy (Figure 30) but it was clear where concentrations were elevated due to biomass burning and fossil fuel combustion. Carbon monoxide was ideal in many ways as a trailblazer satellite measurement. Its atmospheric lifetime against the hydroxyl radical (Chapter 4) is long enough so that there was plenty to observe but short enough that we could observe point sources and hemispheric gradients.

A dedicated satellite instrument called Measurements of Pollution in the Troposphere (MOPITT) developed by the Canadian and US space agencies was launched in 1999 aboard the Terra spacecraft, which also carries other instruments to look at aerosol particles, clouds, and radiation. MOPITT continues to infer atmospheric columns of CO from thermal wavelengths, which are most sensitive to the mid-troposphere. Fifteen years of data have been invaluable in understanding the distribution of CO due to production, destruction, and atmospheric transport.

The next big step of understanding atmospheric composition was by virtue of an instrument intended originally to monitor stratospheric ozone and its precursors, the European Global Ozone Monitoring Instrument. This instrument measured ultraviolet and visible wavelengths (Chapter 2) and was sensitive to a whole range of reactive trace gases that reside in the global troposphere. Scientists from the Harvard Smithsonian and from the University of Bremen were the first to show it was possible to observe weakly absorbing formaldehyde in the measured spectra. Formaldehyde is produced by the oxidation of methane and is present in the atmosphere in very small quantities, but over continents fires and oxidation of biogenic hydrocarbons (Chapter 4) elevate its concentrations substantially. This work together with innovative analysis of satellite observations of nitrogen dioxide at Harvard University and elsewhere to produce global emission maps of nitrogen oxides (Chapter 4) helped to kick-start space-borne measurements of tropospheric chemistry at the turn of the 21st century. With progressively more instruments being launched with higher spatial resolution and higher resolution of a wider spectral region the number and quality of chemistry measurements has increased dramatically. We are rapidly approaching the stage of having a routine capability of observing tropospheric chemistry but, as always, the more we understand the more questions are raised, so our understanding is far from complete. Taking advantage of more and better data scientists are now beginning to link satellite measurements of air pollutants with human health.

A similar revolution is happening with satellite observations of precipitation, clouds, and aerosols. Passive remote sensing techniques have provided us with tropospheric aerosol column distributions on a global scale since the start of the 21st century, allowing us to begin understanding how the size and chemical composition of these particles affect the formation of clouds and how they perturb Earth's radiation balance. With newer, active remote sensing instruments we are beginning to probe through clouds in unprecedented detail to understand the interactions between different sources of aerosols (Chapter 4) and cloud formation.

Rapidly growing atmospheric concentrations of carbon dioxide and methane represent two of the major drivers of contemporary climate change but measuring them from space with near-surface sensitivity represents major scientific and technical challenges. This is because they have long atmospheric lifetimes, so that their concentrations accumulate and any gradients due to local sources and sinks (Chapter 4) and atmospheric transport get progressively smaller. Even the seasonal cycle of carbon dioxide, driven by terrestrial vegetation, represents only a few per cent of the annual mean value. For these data to be useful the data have to be accurate and precise to much less than 1 per cent, making them the most demanding requirements for atmospheric composition. Despite these phenomenal challenges early instruments are beginning to produce data that are of sufficient quality to see expected variations in carbon dioxide and methane.

Solar system exploration

Satellites continue to play a central role in exploring the solar system. There are a number of orbiters currently observing the atmospheres of Mars and Jupiter, and recently the NASA New Horizons did a fly-by of Pluto taking observations of its atmosphere.

The ESA Mars Express mission continues to collect measurements after reaching Mars in 2003. It contains a number of instruments to sample the atmosphere and ionosphere, which have enabled detailed studies, for example, of Martian atmospheric ozone. The

NASA Mars Reconnaissance Orbiter has measured profiles of temperature, pressure, dust, and clouds in the lowest 80 km of the Martian atmosphere since November 2006. This information is now helping to assess potential hazards, such as large-scale dust storms, in support of future robotic and human missions. More recently, the NASA MAVEN (Mars Atmosphere and Volatile Evolution) mission was inserted into a Martian orbit in September 2014 and has collected measurements to understand the slow deterioration of Mars' atmosphere. MAVEN found that this deterioration is greatest during solar storms when streams of charged particles slowly strip away the top of the atmosphere. This process played a role in the the gradual shift from a warm Mars climate with an atmosphere dominated by carbon dioxide, able to sustain liquid water, to the cold, arid planet of today. India's Mars Orbiter Mission also entered a Mars orbit in September 2014 to measure the composition and dynamics of the atmosphere. At only 10 per cent of the price of MAVEN we are beginning to see developing countries making significant headway in space exploration. The European ExoMars Trace Gas Orbiter reached its science orbit in October 2016. The purpose of that mission is focused on characterizing the composition and origin of trace gases in Mars' atmosphere.

Five years after it was launched the solar-powered NASA Juno spacecraft finally reached Jupiter. In July 2016 it began twenty months of collecting data about the Jovian atmosphere, including composition, temperature, and clouds. Juno has already begun to produce stunning images, including atmospheric vortices over Jupiter's South Pole. After that twenty months Juno will slowly descend into the atmosphere.

At the other end of the solar system, nine and half years after launch the New Horizons planetary probe started to approach Pluto in January 2015. Soon after its closest approach to Pluto the probe was aligned such that Pluto atmosphere's was backlit by the Sun. During that time, onboard instruments collected data that showed that Pluto's nitrogen-rich atmosphere extended 1600 km above the plane's surface.

Bringing it all together

Making reliable atmospheric measurements is only the start of the scientific process that is used by weather prediction centres and other scientists. We can learn about the evolving global atmosphere by confronting the best scientific models of the atmosphere with data. Does it agree? Yes, then our model is accurately reproducing the truth. No, perhaps we are learning something new about the atmosphere. However, the sheer volume of data precludes such a simple approach, with some of the fastest supercomputers now necessary to analyse the vast amount of atmospheric data in support of weather forecasting.

Statistical techniques are continually being developed and improved to harness the information embedded in progressively larger volumes of diverse data. Data assimilation describes the process by which we combine information from data and models that embody existing knowledge. This process is not too dissimilar from the least-squares method of fitting a straight line to data, but with a lot more bells and whistles. Despite the size and complexity of real-world applications, the philosophy of the approach remains the same: there is more information from combining data and models than from either the data or the model individually. The mathematics behind data assimilation comes from an unlikely source: a posthumously published paper by the Reverend Thomas Bayes. His idea is conceptually very simple: we modify our prior belief with objective information. The resulting posterior updated belief is a combination of both the prior and the new information. Imagine that: the work of an 18th-century Presbyterian minister informing today's data-driven economy.

Chapter 6
Our future atmosphere

There is still much about Earth's atmosphere we do not fully understand, which limits our ability to predict large-scale changes to the atmosphere. As Earth's climate changes new scientific challenges will emerge that need to be addressed with new measurements and models. These challenges have implications for assessing the impact of future global economic growth and mitigating humanitarian risks. Here, I outline (some) future challenges we face.

(Some) scientific challenges

The atmosphere is changing fast and in many ways in response to changes in natural and anthropogenic-driven climate.

Over the past four decades, measurements have recorded significant changes to the general circulation of the atmosphere (Chapter 3). The Hadley cell transports heat and momentum from the tropics to higher latitudes. The poleward extent of this cell, nominally 30 degrees, helps to define the width of the tropical belt. Atmospheric observations of ozone and temperature and of tropopause heights all point to persistent growth of this tropical belt.

This growth is important because the outer boundary of the tropical belt is where precipitation rates are low, winds are calmer, and where most of the world's deserts are located. History has shown us that persistent changes in the geographical distribution of precipitation, in particular, ultimately lead to large-scale shifts in agricultural and natural ecosystems and to large-scale movement of people, which may have downstream societal and environmental challenges for resettlement. Expansion of the Hadley cell will also change the position of the subtropical jet stream (Chapter 3) and expose new geographical regions to tropical storms. The underlying reasons for this expansion of the tropical belt are still unclear. Climate model studies have highlighted the potential roles of natural climate variability; the increasing atmospheric burden of greenhouse gases, tropospheric ozone, and aerosols; stratospheric ozone depletion; and of changing ocean–atmosphere interactions.

Many countries in the western world have significantly reduced surface air pollution over the past twenty years. These changes have been driven by cleaner fuels, more efficient technologies, and by targeted air quality regulation. But we remain far from living in a clean atmosphere. For example, the UK government in 2013 estimated the UK economic cost of air pollution from human health and the environment. They even broke down this analysis by air pollutant. Particulate matter (Chapter 4), which now affects more people on the planet than any other pollutant, reduces the average life expectancy in the UK by six months, which is worth a staggering £16 billion per year. There are large gradients of pollutants across a city, often determined by many local factors such as traffic patterns, street canyons, and parks. Indeed, a major challenge associated with tackling the urban air quality problem is the increased urbanization of the world population.

Over half of the human population lives in urban areas, an increasing trend that will result in two out of every three people living in urban areas by 2050. The situation is worse in developing countries where there are a larger number of megacities (metropolitan regions with in excess of 10 million inhabitants). A recent report by the Indian government has shown that Delhi has the highest annual particulate matter concentration of any megacity, many times higher than the World Health Organization guidelines. Efforts across the world to improve air quality within cities have had mixed success. Cities such as London have established a congestion charge to enter the city centre. Others have experimented with temporary bans on vehicles, or only allow cars that end with an odd or even number to enter the city centre on alternate days. Without atmospheric measurements it is difficult to determine whether these strategies have succeeded or to develop more effective strategies. Developing sustainable electric transport is a revolution that is already underway, which will also help to reduce air pollution.

We take clouds for granted as we look up at the atmosphere but they play a key role in Earth's radiation budget. Satellites routinely measure the bulk properties of clouds, e.g. coverage, height, thickness. But the properties of clouds and aerosols, and the interactions between them, are arguably the most uncertain components of the climate system (Chapter 2), and the components that could determine the trajectory for future climate. The challenge lies in collecting accurate and consistent knowledge that is necessary to describe cloud processes, which span several orders of magnitude, and representing that knowledge in models.

Unfortunately, at the moment there is a knowledge gap between the detailed microphysical understanding of clouds and improving large-scale climate models that are representative of a few tens of kilometres. Consequently, microphysical cloud processes that are much smaller than climate model grids have to be parameterized, thereby introducing unavoidable errors. One of the main scientific challenges is to seamlessly integrate understanding on microphysical and macrophysical scales, which will require computational and scientific innovations.

But potentially the biggest atmospheric scientific challenge is the one we know nothing about yet; the unknown unknown. These are the emergent or stochastic properties of the climate system that give us little or no warning of their appearance. Two contrasting examples are: volcanic eruptions and emissions of methane from thawing permafrost; there are many more that *could* happen. Volcanoes sometimes give us sufficient warning of their impending eruption to evacuate the surrounding land and air space, and sometimes not. The eruption of Eyjafjallajökull in 2010, for example, highlighted the gaps in our knowledge of the atmospheric transport and chemical transformations associated with lofted volcanic material. The absence of a suitable network of atmospheric measurements led to a conservative restriction on one of the busiest airspaces around and downwind of the volcano. There was also no clear understanding of the boundary between safe/dangerous concentrations of volcanic material.

Boreal ecosystems such as Arctic tundra currently store up to 20 per cent of global soil carbon. Climate models project that surface temperature will increase unevenly over the boreal zone with values higher than the globally averaged amount. Sustained warming will eventually lead to thawing of permafrost, the microbial decomposition of the soils, and to the atmospheric release of methane. Without systematic measurements across the boreal zone it is almost impossible to determine whether the large-scale release of methane from the permafrost has begun.

These examples were chosen for a number of reasons. First, they highlight the spectrum of scales associated with studying changes in the atmosphere, from microphysical to planetary. Second, they focus on a changing atmosphere. While great strides were made in the 20th century to understand the basic properties of the atmosphere, there is still much we do not understand or even know about. In some instances, measurement networks that were established decades ago for one purpose will not be suitable for new, emerging scientific questions that shape our world today. Some would argue in the advent of the satellite era that atmospheric science is transitioning from its early exploratory phase to a more operational phase but, as the examples described earlier show, there is still much we do not understand. These gaps in our understanding are partly due to the atmosphere interacting with the rest of the Earth system. The final reason for choosing these examples is to highlight the links between basic scientific understanding and the associated impacts on humanity and the global economy. Scientific impact cannot be achieved without continued progress in basic science, and likewise basic science is often difficult to justify without some anticipated downstream impact on humanity, even if that impact is seeing further.

The bigger issue is perhaps how, with finite resources, do we prioritize an improved representation of clouds for describing future climate against developing more effective surface air quality mitigation strategies to improve human health?

Technical challenges

Improvements in technology will ultimately help address atmospheric scientific challenges. Sensors are getting smaller and more reliable, data storage is becoming cheaper, and computers are getting faster. But scientific demands will always outstrip affordable technological advances, the equivalent of Lewis Carroll's Red Queen's race in *Through the Looking Glass*.

We are progressively an interconnected global society with smartphones, tablets, and laptops connected via wireless networks—the emergence of the Internet of things. Such devices already include some sensors (e.g. accelerometers) that can be used for environmental monitoring. Integrating atmospheric sensors into smartphones is more difficult if you consider where phones spend most of their time (i.e. in pockets and bags), but other wearable technology might be feasible; in principle any piece of external clothing could be fitted with an atmospheric sensor connected to a network. Sensors could also be installed on the hundreds of thousands of mobile phone masts located across the world or on national energy grid pylons. Literally millions of low-cost autonomous sensors could be installed on buildings and street lamps (using, for example, light-fidelity or li-fi technology) up and down the country, and on cars, trains, and buses. The result would be a map of the near-surface atmosphere on an unprecedented temporal and spatial scale.

Harnessing the power of these low-cost sensors requires their connection to an integrated network, which will include higher specification sensors that help calibrate the sensors. For example, trains with low-cost sensors could get calibrated when they pull into a train station and similarly car sensors could get calibrated when they are off the road to refuel. Imagine traffic and pedestrian routes that could be programmed to avoid congestion *and* minimize air pollution. A smart city is one step further in which traffic signals and public transportation are, for example, interlinked with weather and surface air pollution data. Because everything would then be transmitting information the result would be a data volume that could only be read by machines, so computational algorithms would have to be developed at pace to keep up with the data. Truly a big data challenge.

Space-borne technology necessary to observe higher up in the atmosphere has come a long way since the 1970s. Large-scale missions that require many millions of pounds are still the mainstay of the world's space agencies, but it is certainly not the only option for launching proof-of-concept instruments or a constellation of small satellites. Cube-satellites are created as (potentially interlocking) cubic units ($1U = 10 \times 10 \times 10$ cm^3) and with lower launch costs of tens of thousands of pounds access to space has become affordable for many more groups of people. Developing atmospheric sensor technology to fit in such small volumes is challenging, with inevitable trade-offs between accuracy/precision and size/cost, but worth the investment in effort given the cost. With cube-satellites (and smaller variants), it is now conceivable to launch a constellation of hundreds of low-cost sensors into low-Earth orbit that could be calibrated either with a well-characterized control sensor that is in a similar orbit or on the ground. With the economics of cube-satellites we could potentially launch instruments that have only a few key parameters to measure and begin to monitor the global atmosphere is ways unimaginable in the distant past.

Philosophical challenges

We now know the atmosphere is a component of the Earth system whose properties change with inputs from the Sun and also from Earth's surface. Unlike other components of the Earth system we all share the global atmosphere. It is a global commons. And since no one country has ultimate stewardship of the global atmosphere no one feels ultimately responsible for it.

The continued increase in atmospheric pollutants and greenhouse gases emitted by mankind suggest that the final result may very well be a tragedy of the commons, in which individual national governments make environmental decisions that are sensible on a national scale but do not necessarily make sense on a global scale. Without prompt and decisive collective action involving the world governments this situation will not change.

One proposal put forward to avoid this tragedy, or at least minimize it, is to appeal to market forces: make the global atmosphere and its physical and chemical attributes commodities that can be traded on the world markets. We are beginning to see this happen with carbon, but there are other attributes that can be exploited (e.g. particulate matter). Carbon as a global commodity would, if properly regulated, attract the checks and balances required to progressively minimize carbon emissions. The alternatives are piecemeal efforts that will not achieve success on the scale of the global atmosphere. Because the atmosphere is part of Earth's climate system it is unfortunately a divisive issue. Without an international legally binding agreement on future actions it is difficult to develop effective policy for the atmosphere, placing more emphasis on addressing the scientific and technical challenges.

Further reading

There are some complementary books published by Oxford University Press in their Very Short Introduction series: David C. Catling, *Astrobiology* (2013); Mark Maslin, *Climate* (2013); Tim Lenton, *Earth System Science* (2016); and Storm Dunlop, *Weather* (2017).

David Andrews, *Introduction to Atmospheric Physics* (Cambridge University Press, 2010).

James Holton, *Introduction to Dynamical Meteorology* (Academic Press, 2013).

Daniel Jacob, *Introduction to Atmospheric Chemistry* (Princeton University Press, 2000).

James Kasting, *How to Find a Habitable Planet* (Princeton University Press, 2012).

José P. Peixoto and Abraham H. Oort, *Physics of Climate* (American Institute of Physics, 1992).

John H. Seinfeld and Spyros N. Pandis, *Atmospheric Chemistry and Physics* (Wiley, 2016).

A particularly good overview of why some planets lose their atmosphere and some don't:

David C. Catling and Kevin J. Zahnle, 'The Planetary Air Leak', *Scientific American* 300, (2009), 36–43. doi:10.1038/scientificamerican0509-36

Websites maintained by ESA (http://www.esa.int/ESA), NASA (http://www.nasa.gov), and NOAA (http://www.noaa.gov) are great resources for learning about data, and about past, present, and future Earth and planetary satellite missions.

Index

Index

SOCIAL MEDIA
Very Short Introduction

Join our community

www.oup.com/vsi

- Join us online at the official Very Short Introductions **Facebook** page.
- Access the thoughts and musings of our authors with our online **blog**.
- Sign up for our monthly **e-newsletter** to receive information on all new titles publishing that month.
- Browse the full range of Very Short Introductions online.
- Read **extracts** from the Introductions for free.
- If you are a teacher or lecturer you can order inspection copies quickly and simply via our website.

ONLINE
CATALOGUE
A Very Short Introduction

Our online catalogue is designed to make it easy to find your ideal Very Short Introduction. View the entire collection by subject area, watch author videos, read sample chapters, and download reading guides.

http://global.oup.com/uk/academic/general/vsi_list/

GALAXIES
A Very Short Introduction
John Gribbin

Galaxies are the building blocks of the Universe: standing like islands in space, each is made up of many hundreds of millions of stars in which the chemical elements are made, around which planets form, and where on at least one of those planets intelligent life has emerged. In this *Very Short Introduction*, renowned science writer John Gribbin describes the extraordinary things that astronomers are learning about galaxies, and explains how this can shed light on the origins and structure of the Universe.

www.oup.com/vsi

CANCER
A Very Short Introduction
Nick James

Cancer research is a major economic activity. There are constant improvements in treatment techniques that result in better cure rates and increased quality and quantity of life for those with the disease, yet stories of breakthroughs in a cure for cancer are often in the media. In this *Very Short Introduction* Nick James, founder of the CancerHelp UK website, examines the trends in diagnosis and treatment of the disease, as well as its economic consequences. Asking what cancer is and what causes it, he considers issues surrounding expensive drug development, what can be done to reduce the risk of developing cancer, and the use of complementary and alternative therapies.

DESERTS
A Very Short Introduction
Nick Middleton

Deserts make up a third of the planet's land surface, but if
you picture a desert, what comes to mind? A wasteland? A
drought? A place devoid of all life forms? Deserts are
remarkable places. Typified by drought and extremes of
temperature, they can be harsh and hostile; but many deserts
are also spectacularly beautiful, and on occasion teem with
life. Nick Middleton explores how each desert is unique:
through fantastic life forms, extraordinary scenery, and
ingenious human adaptations. He demonstrates a desert's
immense natural beauty, its rich biodiversity, and uncovers a
long history of successful human occupation. This *Very Short
Introduction* tells you everything you ever wanted to know
about these extraordinary places and captures their
importance in the working of our planet.

www.oup.com/vsi

LANDSCAPES AND GEOMORPHOLOGY

A Very Short Introduction

Andrew Goudie & Heather Viles

Landscapes are all around us, but most of us know very little about how they have developed, what goes on in them, and how they react to changing climates, tectonics and human activities. Examining what landscape is, and how we use a range of ideas and techniques to study it, Andrew Goudie and Heather Viles demonstrate how geomorphologists have built on classic methods pioneered by some great 19th century scientists to examine our Earth. Using examples from around the world, including New Zealand, the Tibetan Plateau, and the deserts of the Middle East, they examine some of the key controls on landscape today such as tectonics and climate, as well as humans and the living world.

www.oup.com/vsi

PLANETS
A Very Short Introduction
David A. Rothery

This *Very Short Introduction* looks deep into space and describes the worlds that make up our Solar System: terrestrial planets, giant planets, dwarf planets and various other objects such as satellites (moons), asteroids and Trans-Neptunian objects. It considers how our knowledge has advanced over the centuries, and how it has expanded at a growing rate in recent years. David A. Rothery gives an overview of the origin, nature, and evolution of our Solar System, including the controversial issues of what qualifies as a planet, and what conditions are required for a planetary body to be habitable by life. He looks at rocky planets and the Moon, giant planets and their satellites, and how the surfaces have been sculpted by geology, weather, and impacts.

"The writing style is exceptionally clear and pricise"

Astronomy Now